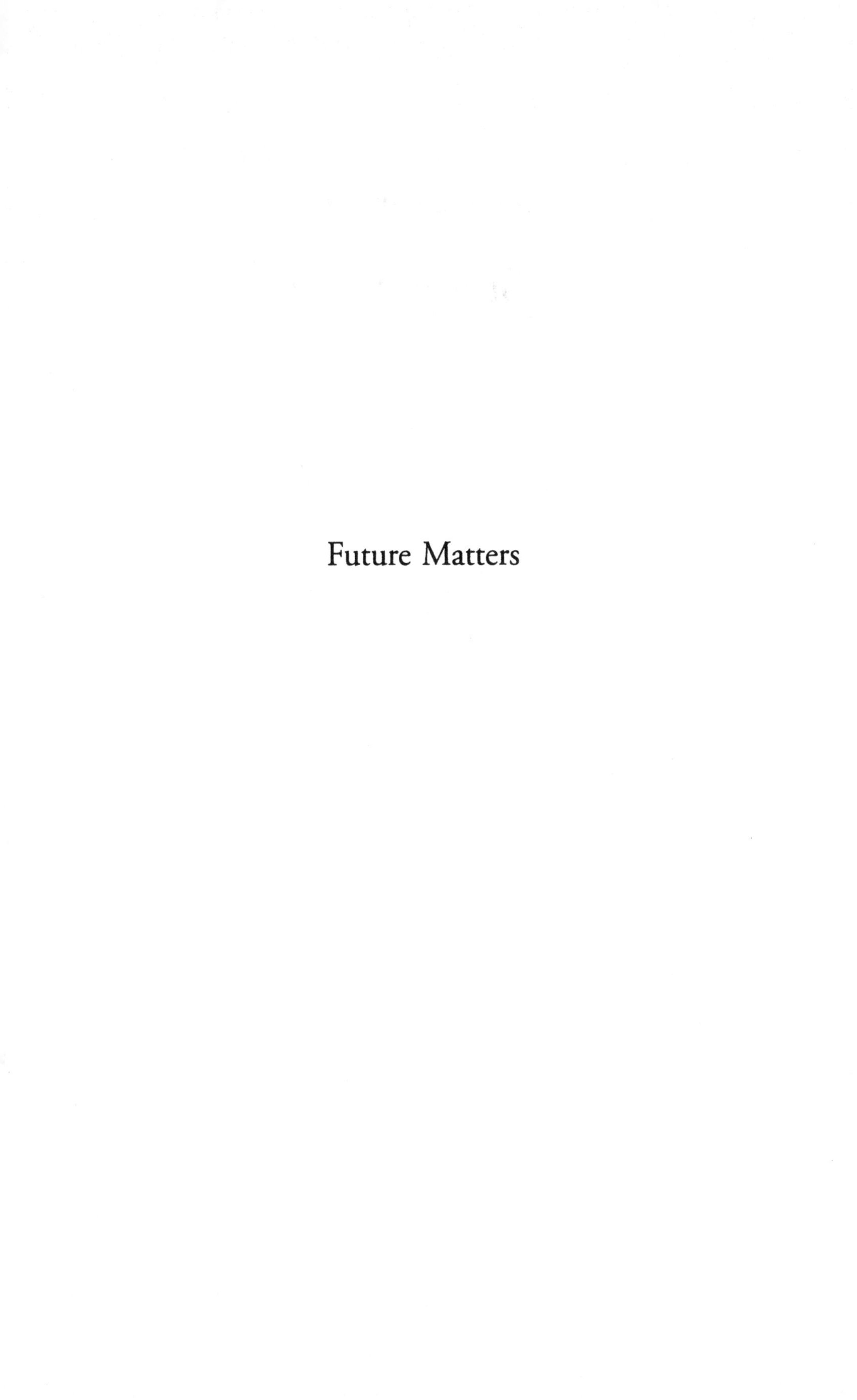

Future Matters

Supplements to

The Study of Time

VOLUME 3

Future Matters

Action, Knowledge, Ethics

By

Barbara Adam and Chris Groves

BRILL

LEIDEN • BOSTON
2007

Cover photo: Paul Wakefield

This book is printed on acid-free paper.

Library of Congress Cataloging-in-Publication Data

A C.I.P. record for this book is available from the Library of Congress

ISSN 1873-7463
ISBN 978 90 04 16177 1

PRINTED IN THE NETHERLANDS

To Joanna Macy
Inspirational Futures Guardian

CONTENTS

LIST OF FIGURES

ACKNOWLEDGEMENTS

This book forms an integral part of a research project on the contemporary relation to the future which has been funded by the UK's Economic and Social Research Council (ESRC) under their prestigious Professorial Fellowship Scheme. It presents the first extended results of three years of research in which the uneven approaches to the future between doing, knowing and caring, that is, between action, knowledge and ethics have been explored. Most of the work produced during this project has been published on the project's website www.cardiff.ac.uk/futures. A number of subscribers to this website have offered themselves as readers of the first draft of the manuscript. We are immensely grateful for their insightful comments, reflections and suggestions for change, some of which have found their way into the final text.

We would like to extend our special gratitude to Jan Adam, Nina Gregory, Gabrielle Ivinson, Ida Sabelis and Ingrid de Saint-Georges who have read the manuscript closely and provided us with many insightful and stimulating suggestions. We would also like to thank others who have offered helpful comments and, in addition, we would like to express a special thankyou to the anonymous referees of the manuscript for their invaluable feedback. Because we have been blessed with so many careful, patient and inspiring readers, it is even more important than is usual for a book such as this that we acknowledge how far the final responsibility for the finished product, and any errors or omissions remaining therein, rests with us alone.

Further, we would like to thank all the members of the project's Review Board for devoting so much of their time and energy to providing us at regular intervals with incisive and painstaking comment on our research, and so many valuable and challenging suggestions regarding its further development.

During the last year and a half, we have also been fortunate enough to benefit from collaborating in a series of workshops with a group of local artists, interested in the theme of futurity, who have engaged enthusiastically with the project's focus and contributed many fascinating insights derived from their own perspectives and practice. Here our thanks go to Philip Babot, Jan Bennett, John Drummond, Alberto Duman and Seth Oliver for their invaluable participation. Finally, we would like to thank all those other participants who attended the project's various international workshops held at Cardiff and the international conference, also entitled *Future Matters*. They have brought to bear on its central themes viewpoints from poetry, story telling, architecture, design, social administration, local government and many other academic and non-academic disciplines. Our research has been enriched immeasurably by their many and varied contributions.

Some portions of Chapter Eight, 'Futures Tended', including the discussion of the applicability of care ethics to future-oriented responsibility, have previously appeared in a truncated form in a paper by Chris Groves entitled 'Technological Futures and Non-Reciprocal Responsibility', published in *The International Journal of the Humanities*, 4(2), pp. 57–62, Winter 2006.

PROLOGUE

Engagement with the future rests on tacit knowledge. We know what it entails and appreciate that it is somehow inextricably bound up with what it means to be human. We learn it as children and, once absorbed, it requires no further teaching or explanation. From a very young age children are expected to be able to employ it. What will you be when you grow up? What do you think you will be when you grow up? What would you like to be when you grow up? We don't think twice about asking children about their expectations and anticipations, their plans, wishes and dreams, unaware ourselves of what might be entailed in such questions: all three of them ask the child to project herself into the realm of the not yet. All three ask her to imagine. Yet they also contain subtle differences: one asks about what *will* be, one what *might* be, while the other still assumes a causal connection between the future and desire, thus implying that futures are made and are there for the taking. Whether we think about tomorrow, the next year, old age or the future of the planet, the difference between probable, possible and preferred futures is likely to play an important role. Whatever the emphasis, however, the future in question here is one that resides in the mind.

Yet, futures are not merely imagined but they are also made. They are produced for months, years and even millennia hence, creating chain reactions that permeate matter and stretch across time and space. These interdependencies, which may not congeal into tangible symptoms for a very long time, make it difficult to anticipate the dispersed potential outcomes of future-creating actions, and so create problems of knowledge. Thus, for example, the innovative use of the earth's resources for the production of energy ushered in the industrial revolution, but it has taken until now for people to recognise the long-term consequences of these practices and begin to accept the need to produce collective responses to mitigate the worst environmental and climatic impacts. Not our generation, however, but an open-ended line of successors will have to endure, absorb and deal with the long-term effects of developments we largely associate with scientific and technological progress. The difficulties associated with knowledge about outcomes of actions, in turn, raise uncomfortable questions about responsibility.

Future Matters addresses this complex relation between action, knowledge and ethics. Wherever we care to look we cannot fail to notice that contemporary industrial societies' capacity and competence to produce futures is phenomenal. These created futures potentially reach to the end of time. In contrast, knowledge of such futures is dismal. It cannot encompass the potential reach of socio-technical actions and their effects. And because the future cannot be known, responsibility tends to be pushed outside the frame of reference and concern. In Future Matters we begin to investigate this uneven relation between acting, knowing and taking responsibility. We ask: 'How did we as a society get here?' 'Have we always related to the future in this way?' 'If not, what was different and what might be learnt from these other ways of engaging with the not yet?' 'What would need to change for the relation between action, knowledge and ethics to become less disparate and more in tune with the temporal reach of contemporary future making?'

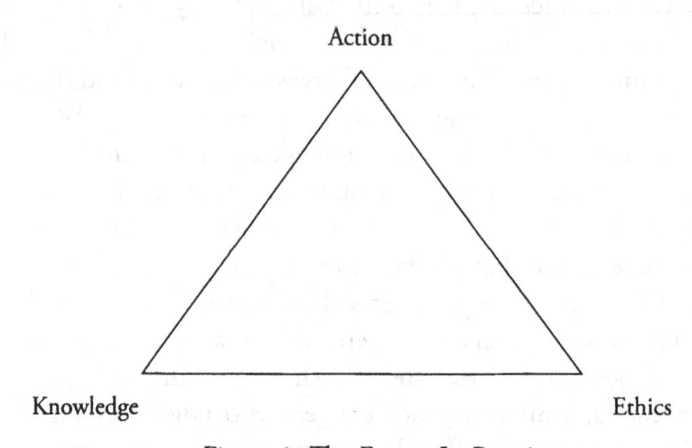

Figure 1: The Future In Practice

The search for answers to these and related questions has taken us on a wide-ranging journey into often unfamiliar and unexplored territory. Along the way we have encountered many surprises and gained insights that were difficult to achieve but which we would not want to have missed. We have met fellow travellers from all walks of professional life who shared our passion to find better ways to engage with the future. Sometimes our paths merely crossed and we encountered each other with friendly interest. At other times we walked together for a little while before we once more went our separate ways as our respective concerns pulled us in different directions. Some fellow explorers have been along these paths many years ago and have left markers and cairns for us to follow. Others, we are convinced, will be following. For these, and for fellow citizens who have not yet realized that they share this problem, we have written this book. For them we marked and mapped relevant tracks (especially the passages through virgin territory) and where we could not go any further we left clues as to where the paths might lead.

In the course of our research we have found new interdependencies and connections and began to recognise issues as significant that hitherto we thought unproblematic and trivial. Most importantly, we struggled to develop concepts that could express and explicate what we know and live at a subliminal and intuitive level. To help the reader with some of these key terms we have developed a glossary. Each time a glossary term first appears in a chapter, it is marked with an asterisk (*). In addition we sought ways to disrupt our world of taken-for-granted assumptions so that their empowering and delimiting role in the domain of social action could become recognised. This we considered an essential pre-condition to effecting change. Finally, we had to appreciate that engagement with the future is an encounter with a non-tangible and invisible world that nevertheless has real and material consequences. Making the invisible visible and tangible, therefore, was a further task we had set ourselves for this book. We are no strangers to this problem of invisibility as our sphere of expertise is social time. The temporal world is fundamentally intangible, and is accessible to the senses only through its material results: the grey hair of friends we have not seen for many years, the geological strata exposed after a rock fall, the cancers of the children and young adults who live in the fall-out zone of the Chernobyl nuclear explosion. Here we focus on one temporal domain, the future, and bring to bear on it our understanding and ways of seeing. The work presented in these pages therefore is informed by and grows out of our work on time.

Writing the book we were propelled along by a sense of urgency. The issues we were addressing are pressing, requiring socio-political action now or in the very near future. Despite this, however, the book is not programmatic. It does not set out a blueprint for action in response to the problems we identify. Instead, it opens up spaces for collective questioning and debate, provides opportunities to conceive of alternatives. By painting with a broad brush on a vast historical canvas of cultural history that reaches to Greek antiquity and beyond, we show that things have been done and thus could again be dealt with differently. While not programmatic, the book is unashamedly normative. As authors we accept ourselves as participants and contributors to future making rather than observers of such practices. What we tell and how we tell it is not an objective representation of the world as it is. It is always already and inescapably selective. It prioritises some issues over others. It gives more credence to one approach than another. It foregrounds what we consider significant and by implication ignores, thus silences, other domains of action, knowledge and ethics.

The introductory chapter sets out the problem, maps the parameters of the issues to be discussed, provides insights into our approach, draws attention to our concerns and identifies the stations and staging posts to be visited. As a starting point for the exploration the introduction merely senses some of the connections and interdependencies that are explored and developed in the book. From there the story is told sequentially and builds up as the chapters progress. Thus, while each single chapter can stand on its own, the analysis of the whole will only reveal itself after all of the chapters have been read in their designated order.

Throughout the book, we are interested in the relationship between worldviews inspired by modern science and their influence on contemporary societies' attitudes to the future. At various points, we refer to the worldview of 'mechanistic science' and how it limits the scope of our capacity to engage with the future. We would like to emphasise at the outset that, in pointing out the limitations of this worldview, we are not criticising science *per se* as mechanistic, overly deterministic etc. We recognise that, for scientists, the relationship between past, present and future is perhaps not so clear cut as the mechanistic worldview we describe. Particularly where science is concerned with the quantum level of reality, or complex systems, this is not the case. Our focus in this book is on how scientific research is *used* as the basis for evidence-based public policy. In its applications, all too often the nuances of scientific theory and explanation are lost in the rush to establish reductive causal explanations and develop technological applications. When we refer to a worldview associated with 'mechanistic science', it is this we have in mind.

Finally, a brief note on our use and choice of pronouns may be useful at this point. We have decided to vary our use of the masculine and feminine pronouns without special regard to context. Our use of 'we' expands and contracts according to context and is rarely specified in its precise delimitation. First, it encompasses the authors. Widening the circle, the next level of 'we' includes the reader. Since a book of this nature is very much a joint production between authors and readers we are not unduly worried about a few ambiguities and possible confusions between those two kinds of 'we'. The next expansion of 'we' takes in members of contemporary industrial societies and from there works backwards to predecessors with whom we share a cultural history. On a few rare occasions the 'we' is extended to all of humanity and all fellow living beings. Our choice of 'we' therefore covers primarily the western tradition of thought. This is based on an explicit and conscious decision. Let us explain. The relations and approaches to the future we are discussing in these pages are ones which are inextricably linked to the industrial way of life, along with the problems and (ir)responsibilities that accompany them. It is therefore from within this tradition of knowledge practices that change has to emanate, that the seeds for change have to be gathered, sown and tended.

In Future Matters we begin this process. We draw on the collective cultural history of the west to connect our contemporary ways with those of predecessors who found different means of taming the unknown. We show that things were and therefore could again be different. In the course of the exploration we raise many questions, establish connections across disparate knowledge spheres, disrupt the *status quo* and pursue our commitment to identify openings for change so that alternative practices may flourish. We very much hope that we can infect you with our enthusiasm for this subject and that you too are encouraged by the potential for change it opens up.

Barbara Adam and Chris Groves
Cardiff, March 2007

CHAPTER ONE

INTRODUCTION

Industrial capitalist societies are inescapably wedded to innovation and progress*. Change rather than stability is the order of the day. In this dynamic world of universal mobility, standing still means falling behind. This committed pursuit of novelty distinguishes this contemporary mode of being, so aspired to across the world, from other socio-economic systems in which the creation of permanence and stability was and is the desired goal, where products were and are crafted to last, where political structures are designed to endure and people conduct their social relations with a fair measure of predictability. The degree to which societies actively seek change, or permanence, has significant implications for their relation to the future. As Bertrand de Jouvenel explains,

> The fewer changes we anticipate, the more we can continue to rely on our knowledge for the future. If society tends on the whole to conserve the present state of affairs, our present knowledge has a high chance of being valid in the future. On the other hand, the future validity of our knowledge becomes increasingly doubtful as the mood of society inclines toward change and the changes promise to be rapid. (de Jouvenel 1967: 10)

In *Future Matters* we seek to contextualise present efforts to anticipate and traverse the future within the wider history in Western culture of telling, taming, trading* and transforming the realm beyond the present, which extends back some 5000 years into antiquity.

Contemporary societies dedicated to progress, innovation and change, we want to argue, need to hone their tools for anticipating, taming and transforming their futures. Since the pursuit of change radically reduces stability and with it structural security, the substantial effort required to achieve competence in futurity is the price to be paid for the prize of advancement on all fronts of knowledge and socio-economic growth that awaits those most committed to the system of accelerating change. In their "Communist Manifesto", Marx and Engels describe the relation between the pursuit of progress and the production of uncertainty in the following way.

> Constant revolutionizing of production, uninterrupted disturbance of all social relations, everlasting uncertainties and agitation, distinguish the

> bourgeois epoch from all earlier times. All fixed, fast-frozen relationships, with their train of venerable ideas and opinions, are swept away, all new-formed ones become obsolete before they can ossify. All that is solid melts into air, all which is holy is profaned, and men at last are forced to face with sober senses the real conditions of their lives and their relations with their fellow men. (Marx and Engels 1967/1848: 224)

Given that social uncertainties and insecurities increase proportional to the efforts in economic and technological innovation, greater social transience needs to be counterbalanced by a parallel increase in concern with the future. In *Future Matters* we take this task seriously by exploring some of the contemporary ways of handling the future, considering their adequacy and identifying some openings for departure from the established traditions.

This introductory chapter to *Future Matters* provides the structural framework for the historical chapters of this book. In five sections it outlines how the future has been told, tamed, traded and transformed and how it is traversed today in a way that is superimposed on those earlier relations. It distinguishes the embedded, embodied, contextual future from contemporary perspectives on a decontextualised future emptied of content, which is open* to exploration and exploitation, calculation and control. It shows how the *emptying** of the future is implicated in both the progress of industrial-capitalist societies and the major problems that they face today. In the concluding section it suggests that there is much to learn from the conceptual tools honed by predecessors in their efforts to render the future more tangible. The structure of this introductory chapter therefore foreshadows the historical part of the book, which is followed by a conceptual critique of deeply embedded habits of mind* and an attempt to identify access points for change.

The Future Told

The desire to unlock the secrets of fate and make contact with the realm beyond the present is shared by archaic and modern cultures alike. Throughout the ages, this quest has taken numerous forms and has been entrusted to many different kinds of gifted specialists. In his history of divination*, John Cohen (1964) gives accounts of over one hundred ways of telling the future. From the history of prophecy* and divination we learn that foreknowledge of the future was the prerogative of gods, the gift of prophets, oracles and seers, of witches and wizards and, more recently, the specialised task of astrologers. Each of these specialists drew on different sources of privileged knowledge and used different methods to access the temporal realm beyond the senses. A few examples will serve to illustrate the diverse ways of unveiling a future that was first conceived as pre-existing destiny and only in more recent times, as evolutionary continuity with the past.

The art of prophecy is believed to have originated some 5000 years ago in Mesopotamia, the land between the rivers Tigris and Euphrates, where some of the earliest records point to a range of prophetic and divinatory methods. The oldest prophecies appear to have been dream interpretations. According to Richard Lewinsohn (1961: 51–7) the peoples of Mesopotamia considered the legendary Sumerian king Emmendurana (or Enmeduranki), whose reign predated the great flood, as the founder of dream divination. Later methods of foreknowledge were based on hepatoscopy, the inspection of the liver. The liver was considered the seat of life and interpretation of its symbolic characteristics was the prerogative of respected specialists who were answerable to their kings and nobles. Clay models of the liver, discovered by archaeologists, appear to have been used to instruct future prophets in their important art. In ancient Sumeria and Mesopotamia, as in cultures that absorbed and adapted some of these very early ways of engaging with the unknown, gods and god kings were the source of that prophetic knowledge. Ancient Egyptian societies too greatly depended on prophecy for guidance on prospective action and relied on it to be forewarned of impending disaster. From archaeological records we learn that Egyptian oracles were primarily connected to festivals and associated with temples. It was in this way and at these specific times and places, Jan Assmann (2001/1984: 35) suggests, that "the city deities exercised their de facto rulership", which "reached beyond the temple enclosures and included the entire citizenry."

Still in this part of the ancient world, the Israelites' relation to the future is recorded in their sacred texts. Thus, the Old Testament is replete with stories about prophecies, visions, revelations and dreams through which Jehovah revealed his will. For example, eighteen of the thirty-nine books of the Old Testament carry the subtitle 'The Book of the Prophet', telling of things that came to pass, of prophets employed to guide the Israelites on their way to freedom, by prophets acting as conduits for God's messages. In Genesis (6.13 onwards) God speaks directly to Noah, warns him of the impending flood that will destroy all of creation and instructs him to build an ark in which he is to save his family and the animals, two of every kind. By following God's instructions, Noah was able to escape the flood. Similarly, Moses was able to lead his people to the chosen land. All significant elements of Christ's life too are prophesied in the Old Testament, while prophecies of the end of the world and Judgement Day predominate in the New Testament. Thus, the prophets of the Bible, with their privileged access to a time that is inaccessible to the senses, were depended upon extensively to provide guidance and forewarning, signal and council. Their foreknowledge was based not on their own clairvoyance or wisdom but was imparted to them by their God.

What distinguishes Greek oracles from the prophets of ancient Sumeria, Egypt, and the Israelites is the nature of the oracle's prophetic gift. While Greek gods too had a role to play in the oracle's prophecies, their input was considered less reliable, their gods' characters as well as their actions being much more like those of humans than the gods of the cultures discussed above. Moreover, the Greek oracle's prophetic prowess was associated with special hallucinogenic powers. From ancient Greek mythology we know that few decisions of significance were taken without first seeking advice from oracles, the most famous of which was located in Delphi, the place that Zeus had declared the centre of the universe. The stories of Perseus and the Gorgon, of Hercules and of Oedipus all tell of attempts to avert the destiny prophesied by the oracle. In each instance, however, the prophecy came to pass: fate proved stronger than diverting actions taken in the light of foreknowledge provided by the oracle. For Cassandra, a tragic figure in Greek mythology, the gift of clairvoyance was a curse: no one would listen to her warnings or believe her prophecies. Thus, in vain she forewarned of the fall of Troy and the trick with the wooden horse. She even foresaw the details of her own and her husband Agamemnon's murder. Yet she was unable to avoid her destiny, helpless in the face of her own foreseen demise. In all these stories the message is clear: knowing the future does not necessarily help you to alter your destiny. Fate is pre-set and resists our best efforts to influence its course. In Nordic myths too, we find a strong tendency for prophesies to come true no matter how hard gods and mortals try to avert the destiny thus prophesied. For example, in the myth of 'The Death of Balder the Beautiful' (Ferguson 2000: 38–43) even Odin, the all-powerful god of light, and his wife Frigg are unable to prevent the killing of their son, as foretold.

The Druids of ancient Celtic cultures, who were equally renowned for their powers of divination, drew on different sources and powers. As Danah Zohar (1983: 16) notes, Celtic Druids read the future from "the flights of birds, from the shape of clouds or tree roots, with the aid of bone-divining (using the boiled-clean right shoulder blade of an animal) or from rowan sticks". Merlin, hailed as the greatest wizard of all, foresaw many of the significant turning points in the life of King Arthur and far beyond that. His prophecies and predictions cover the crusades as well as the reign of James I, Henry VIII and Richard the Lionheart who lived some 700 years later. Thus, Merlin prophesied that

> The Lionheart will against the Saracens rise,
> And purchase from him many a glorious prize...
> But whilst abroad these great acts shall be done,
> All things at home shall to disorder run.
> Coop'ed up and cage'd the Lion then shall be,
> But after suffrance ransom'd and set free.
> ...Last by a poisonous shaft, the Lion die. (quoted in Zohar 1983: 18)

In their foreknowledge of the future, Merlin's fellow druids were said to have been similarly accomplished, performing their task without recourse to hallucinogenic or hypnotic aids.

The reading of patterns, recognising significant coincidences, understanding synchronicity and establishing a-causal connections to unlock the future are in fact the means that are shared across history by people with special access to the future: by prophets and oracles as well as Druids and Nordic sages. The principles upon which these divinations are based, we need to appreciate further, are diametrically opposed to those underpinning mechanistic scientific prediction, the primary, dominant and socially most legitimated means within modern industrialised cultures of accessing the realm beyond the senses. In contrast to these ancient practices, scientific prediction is wedded to the principle of linear causality and projects the repetition of past patterns into the future. Historically, the rise of scientific prophecy is interesting. Despite the fact that scientists were trespassing on a terrain that was the exclusive preserve of God, scientific prediction gained acceptance from the Christian church on the basis that it merely brought together knowledge about processes that had occurred in the past and were therefore expected to continue into the future. That is to say, given that the past rather than the future was the source of science's prophetic prowess, the church did not consider predictive science to be either blasphemous or the work of Satan.

Before we can focus on knowledge practices* that make social life more predictable and thus more secure, we need first to appreciate that the futures outlined above are tied to the assumption that 'the future' is pre-existing, that there already *is* a future to be discovered and told. Only on the basis of such pre-existence and pre-determination can the future be unveiled and can we make sense of efforts to intervene in fate and destiny.

The Future Tamed

Efforts to know what lies ahead have to be distinguished from knowledge practices that make daily life less precarious. These latter practices are concerned not so much with knowing and intercepting destiny but more with providing structural security for the daily and seasonal rounds of social life. Such structural security can be established by better anticipation of natural rhythms, social interactions or both. It entails a quest for special skills and know-how associated with futurity. Know-how implies knowledge for use, that is, knowledge to structure, order and tame the insecurities of the realm beyond experience. It encompasses knowledge which is useful in efforts to render the uncertain more certain, the insecure more secure, and the unknowable more knowable.

For example, ancient Egyptian death rituals facilitated the detailed anticipation of life after death. Texts written on the walls of tombs, on clay tablets and much later on papyrus, provided authoritative information about what to expect after death and on how to behave to ensure a safe journey to the netherworld (Assmann 2001/1984; Geddes and Grosset 1997; Hornung 1999/1997). In ancient Egyptian society death was seen as a key marker in the stream of existence, a difficult staging post beset by perils and unforeseen hazards. However, by following the examples of gods and especially Osiris, the rituals systematically transformed the abyss of the great unknown into something familiar and unthreatening. Thus, the "Book of the Dead" is a book of spells whose sole purpose is to ease the journey of the dead person to the afterlife. The "Book of what is in the Underworld", in contrast, describes the underworld, thus taking away some of the fear of the future unknown. It exists in many versions and has been found in the tombs of both kings and ordinary citizens. (While in the old kingdom the transfer to eternity was the preserve of pharaohs, in later times it was open to anyone who could afford the rituals necessary for safe passage.) The 'Pyramid Texts', finally, are intended to ensure entry of the dead to the netherworld. They mostly recount the Osirian legend and give detailed guidance on how to emulate Osiris' transition from the world of the living to the realm of dead souls. Underpinning all these instructions was a belief that the preserved physical body was essential to securing existence in the afterlife. From these death rituals of ancient Egypt we can see how detailed 'knowledge of the unknown', in this case the stages of death and the journey to the realm of the dead, provided existential security. Clear instructions for rituals relating to the deceased, therefore, transformed ministering to the dead from an anxiety-bearing last service into an essential life-giving activity. Non-existence, the ultimate unknown, had been rendered knowable. From the secure basis of practical knowledge, therefore, the future in the netherworld of dead souls became a mere technical challenge and a matter of correct ritual conduct.

In a less technical way, religion in the most general sense fulfils the need to know about the unknown, the life beyond death, the world beyond human existence. It places human beings in the wider scheme of things: nature, the cosmos and the spirit world. It explains continuity and locates every person's finite life in the greater cycle of life, death and renewal. The details may differ between the world's religions but each one provides a measure of predictability about the great unknown, life after death. Each one offers guidance about conduct and the consequences of actions. Each one tames the unknown future, renders it knowable and known.

Beyond ritual and religious creation of existential security there are numerous other ways in which practical knowledge has been able to enhance the predictability of the realm beyond the present and thus increase the structural security of social existence. These may relate to knowledge about the movements of planets and their impacts on seasonal and climatic patterns, they may entail the creation of institutional structures and they may involve postulating connections between planetary patterns and social destiny, as in the case of astrology. Repeating cycles allow for the recognition of patterns. Naming and numbering these repetitions makes them predictable, allows for anticipation and planning.

From the studies of archaeoastronomy in Britain (Ruggles 1994), South America (Aveni ed. 1975) and the Middle East (Heggie ed. 1982) we know that buildings were aligned with the stars so as to bring into unity heaven and earth, social organisation and the divine. The rising and setting of heavenly bodies was tracked and fixed against features in the landscape while the key features of buildings were aligned with respect to the local horizon and the extreme positions of sun, moon and planets. Thus, solstice and equinox, the moon cycle extremes (which repeat every 18.61 years) as well as the disappearance and reappearance of stars have all been connected to aspects of ancient built structures as distant in time and space as the pyramids and temples of ancient Egypt, the temple structures of Inca, Maya and Aztec cultures, and the stone circles, long barrows and cromlechs of Neolithic Britain.

As Adam argues elsewhere (Adam 2004: 102–12), in each of these cases, the knowledge of repeating cycles enhanced anticipation and facilitated social activity in preparation of future events. When such natural rhythmic processes are integrated with social regularities of seasonal activity and religious festivities they help to anchor social life in patterns of anticipated events, thus tame what would otherwise have remained key insecurities of social existence. Moreover, such practices demonstrate that knowledge of the future is not just possible but is an essential ingredient of social life.

In addition, de Jouvenel (1967) points to the importance of social certainties as pre-conditions to any form of socio-cultural life. These certainties, he suggests, are created on the basis of expected social behaviour.

> They may be interpreted as an offensive collectively waged on the future and designed to partly tame it. As a consequence the future is known not through the guesswork of the mind, but through social efforts, more or less conscious, to cast "jetties" out from an established order and into the uncertainty ahead. The network of reciprocal commitments traps the future and moderates its mobility. All this tends to reduce the uncertainty. (de Jouvenel 1967: 45)

Habits, customs and traditions as well as laws, rules and moral codes provide a degree of foreknowledge and anticipation. They make the behaviour of others predictable and facilitate a certain measure of security. In a similar vein, contracts and promises, obligations and commitments allow for projective actions to be embedded in socially constituted frameworks of certainty, thus making possible calculations and estimations for what would otherwise be unpredictable transactions. Structural and contractual securities thus form additional stable anchorage points in the shifting sands of the social future and they allow for a wide range of predictions with a high degree of probability. In this instance it is the use of social institutions and practices through which the future is tamed and uncertainty reigned in sufficiently to make social interaction possible.

Irrespective of whether it is tamed by ritual, religious, astronomic or institutional means, the future delineated above is one of embodied and embedded continuity. It is contextual and personal and as such located in wider systems and structures: social, natural, cosmic and spiritual. To tame this future is to know and understand the wider scheme, rhythms and processes within which individual lives are embedded. Once these are known and understood, practical action can be taken to counterbalance the terror of non-existence and the impending unknown.

Futures Traded

The trade in futures originated from rather similar concerns. It was initially an attempt to create greater security within a context of uneven fortunes in commerce and over a person's lifetime. It addressed the great variations within and between seasons as well as longer time frames that affected livelihoods. Insurance, banking and the trade in 'futures' were eventually born of these intentions, which appear at first sight to be no different from the knowledge practices associated with the taming of the future. On closer inspection, however, we find that a fundamental shift in perspective and assumption about the future had taken place. Without this fundamental shift in understanding and approach the trade in futures would have remained a sin. The Christian church was quite clear on this, and so is the Koran: it was a sin to trade for profit something that belonged not to human beings but to God/Allah. According to the doctrines of these religions, trade in futures was/is theft because it trades in something that could not belong to individuals.

As long as earnings on the future were deemed to be a sin, explains the historian Jacques Le Goff (1980), capitalism and the money economy could not develop since, for the merchant, time was one of the prime opportunities for profit. There could be no charging of interest, no trading or discounting of futures*. The merchant's activity is based on assumptions of which time is the very foundation—storage in anticipation of famine, purchase for resale when the time is ripe, as determined by knowledge of economic conjunctions and the constants of the market in commodities and money—knowledge that implies the existence of an information network and the employment of couriers. Against the merchant's time the church sets up its own time, which is supposed to belong to god alone and which cannot be an object of lucre.

While interest and credit had been known and documented since 3000 years BC in Babylonia, it was not until the Middle Ages that the Christian church slowly and almost surreptitiously changed its position on usury (Le Goff 1980: 29–100; Wendorff 1991: 131–146), which set the future free for trade, to be allocated, sold and controlled. For Muslims, in contrast, the ruling still applies.

Since the late Middle Ages, trade fairs existed where the trade in futures became commonplace and calculations about future prices an integral part of commerce. The trade in futures buys and sells futures for the benefit of the present, that is, for profit in the here and now. It trades not just in goods but the *promise* of goods. This was of particular importance for international trade by sea given that trade ships might be away for as long as three years at a time. It involved global merchants in complex calculations about potential profit and loss over long periods, with goods being traded in absentia, thus establishing the trade in futures as an integral part of the western, capitalist economic system. While banks calculated the monetary value of the future with respect to interest and credit, insurance companies calculated it with respect to future risk. For regular payment the insurer promised to compensate for potential loss and disaster thus helping to smooth out fluctuations in personal and corporate fortunes. The development of statistics during the second half of the seventeenth century dramatically improved such calculations, affording glimpses of futures with high degrees of probability* (Lewinsohn 1961: 87–9).

In all these economic strategies the future is commodified*. At issue is no longer the embedded, contextual, embodied future discussed earlier but a future emptied of content and divorced from context, a future that can be calculated anywhere, at any time and exploited for any circumstance. This commodified future is irreducibly tied to clock time and its economic use as an abstract exchange value. The difference between empty* and contextualised futures is of significance. Embodied futures, we need to understand, cannot be traded. Just as the planet Pluto's trajectory cannot be exchanged for that of Saturn so my future is not exchangeable with that of the oak tree in my garden. The commodified future, emptied of all contents, in contrast, can be calculated, traded, exchanged and discounted without limit. This difference between empty and embodied, contextual futures is one that will concern us for much of the remainder of this introduction.

Futures Transformed

The effort to intervene in the future, as we have shown above, can be traced back for some 5000 years. It can be understood as an attempt to change pre-existing destiny. Contemporary endeavours to transform the future, in contrast, are far more ambiguous than earlier attempts at intervention. The contemporary idea of transforming the future carries within it a much stronger element of human influence as well as an underlying assumption that the future can be shaped according to human will. At the same time, it retains the notion that there exists something which is to be transformed: the world of nature, evolution, genetic inheritance, for example. Where the emphasis is placed along the continuum from intervention to transformation and creation depends on whether or not the future is embodied and embedded in processes* and events or decontextualised and emptied of content. A commodified future, as we have indicated above, is neither tied to destiny nor conceived as pre-existing. Rather, it is an open future, a realm of potentiality to be formed rather than transformed to human will. Emptied of content and meaning, the future is simply there, an empty space waiting to be filled with our desire, to be shaped, traded or formed according to rational plans and blueprints, holding out the promise that it can be what we want it to be.

Many of the inconsistencies we find in contemporary writings on the future can be traced to ambiguity on this issue, that is, to the fact that the difference between the principal assumptions about the future are left implicit, rarely thought about or theorised. The quest for knowledge about an embedded, embodied, contextual future and later efforts to tame that realm of existence, we want to argue, need to be understood in relation to and in distinction from socio-economic activities that seek to shape, form and colonise a future of our own making. In addition, the pursuit of knowledge about the realm(s) beyond the present needs to be distinguished from the *creation* of the future. Similarly, the contemporary conjecture *about* future events needs to be differentiated from cultural forays *into* the future. This point will become clearer when we explore the idea of the modern future being traversed. Before we move to this issue, however, we need to briefly address the issue of increasing uncertainty.

There seems to be agreement amongst scholars concerned with the future that contemporary futures are marked by far greater uncertainties than were encountered in traditional societies. This rise in uncertainty is associated, for example, with the pursuit of progress and accelerating change, with the reduction in structural certainties and with the increase in mobility of just about everything: people, objects and information (Brown, Rappert and Webster 2000; Lewinsohn 1961; de Jouvenel 1967; Ling, 2000; Nelis 2000). As we noted at the beginning of this chapter, Marx and Engels came to a similar conclusion some 150 years ago on the basis of their detailed analysis of the emerging capitalist system.

Today there is a need to extend that analysis further and develop it in different directions. We mention here just two examples. To better understand the growing uncertainty observed in contemporary societies we have to encompass the role of information and communication technologies (ICTs) on the one hand and to take account of the decontextualisation of the future on the other. These issues will be addressed in later chapters of the book. All that can be done here is to mention some of the key points in need of consideration. Regarding the relation between ICT and the rise in uncertainty, the following developments seem to be of importance. With ICT, succession and duration have been replaced by seeming instantaneity* and simultaneity*. That is to say, duration has been compressed to zero and the present extended spatially to encircle the earth. For people with access to ICT, and those implicated in their effects, therefore, the present has been globalised and, as Paul Virilio (1997) puts it, intensity has taken over from extensity, bringing with it the possibility of concerted action in 'real time'. The crucial element is not, however, the vastly increased speeds involved but, rather, the networked nature of information that is distributed simultaneously across space. It is the loss of causality and sequence, rather than the associated speed and acceleration alone that so dramatically inclines this system of information transfer towards indeterminacy. When in principle everyone has access anywhere with the potential to influence anyone, certainty is no longer attainable (Hassan 2003).

Equally important for the rise in uncertainty, we want to propose, is the abstraction of the future from its embodied and embedded position in socio-economic, political and socio-environmental processes and events. Once emptied, the future can be filled with anything, with unlimited interests, desires, projections, values, beliefs, ethical concerns, business ventures, political ambitions.... It becomes a free-for-all, unbounded, unlimited and thus fundamentally and irreducibly indeterminate. In contrast to the context-bound future, the empty future of contemporary economic and political exchange is fundamentally uncertain and unknowable. At the same time, however, it appears wide open to colonisation and traversal.

Futures Traversed

The future, emptied of content and extracted from historical context, invites imagination and inventive action. It is ready to be populated with the products of progress. An empty future is there for the taking, open to commodification, colonisation and control, available for exploitation, exploration and elimination, as and when it becomes appropriate from the vantage point of the present. We would like to go as far as to suggest that the emptying of the future and its subsequent equation with money were central preconditions to the progress enjoyed by industrial societies, to the economic growth experienced by those societies and to their colonial ventures. When the future is decontextualised and depersonalised we can use and abuse it without feeling guilt or remorse. We can plunder and pollute it with impunity. We can forget that our future is the present of others and pretend that it is ours to do with as we please, with our imagination, creative skills and technological prowess the only boundaries to our activities. This is the base assumption upon which our present affluence and domination has been created and which we carry forward to 'our' future. Today, however, it becomes ever more difficult to keep up the pretence.

The fiction of an empty future, which took us to the height of economic wealth and global domination, is starting to disintegrate, the embodied future rising from its suppressed position. We are beginning to recognise that our own present is our predecessors' empty and open future: their dreams, desires and discoveries, their imaginations, innovations and impositions, their creations. Our progress and our pollution, our colonial and contractual responsibilities as well as our global institutions, markets and finance are their empty, open, commodified futures in progress, are their creative imaginations working themselves out *in* and *as* our embodied and embedded present. We are the recipients of their pretence, their illusion, which is for us inescapably real in its consequences.

Our present is their created future, their commodified future and their colonised future. Our present is their uncertain future, where 'all that is solid melts into air'. Our present is their discounted future, their future which was exploited commercially for the exclusive benefit of their present. Our memorials are their political aspirations, their ethnic cleansing. Today we still pursue that same illusion, still live the same pretence. The future is open, we say. We cannot know it. It is open only to our imagination. It is ours to forge and to shape to our will, ours to colonise with treasured belief systems and techno-scientific products, ours to traverse, ours for the taking.

As long as everyone colludes, the house of cards stays intact. In a globally connected and interdependent world, however, not everyone is willing to play the game of 'let's pretend'. Not everyone is willing to concede all of the world's treasures, present and future, to the player with the strongest card. From across the world, competing claims on the future are making themselves heard. 'Others' want their fair share of the potential bounty, or want very different things. Meanwhile, claims are being filed for the results of predecessors' faith in empty futures. Accusations accumulate about past wilful blindness. Predecessors' glorious creations rebound as nightmares. The costs have to be paid, the disasters rectified, the cancers endured. Despite mounting evidence against the belief in the empty and open future, however, this perspective shows little signs of getting weaker. Against strong countervailing pressures it retains its dominant position. It still gathers strength through political rhetoric, scientific promise and the quest for economic wealth. It is spread and perpetuated by management gurus. The empty and open future unites them in a common belief, while in the meantime embodied, historically embedded, contextual futures of processes and productions are airbrushed from the picture, traversed and negated.

Today, this continuing negation requires our most serious attention and its transcendence is the urgent and important task facing scholars and practitioners who have come to recognize the global effects of the elimination of embodied futures from the frame of reference. At the level of scholarly engagement this requires a shift in perspective and focus. It demands historical perceptiveness, asks for a thorough knowledge of temporal relations and calls for a trans-disciplinary outlook. At the practical level it necessitates compassion with an eye for justice and an acute awareness of the interconnectedness, interdependence and interrelatedness of everything. As such it calls for conceptual skills and practical tools similar to those that ancient societies had honed to perfection: to understand processes and events in the wider scheme of things, to recognise connections and implications, to appreciate things in their continuity and emergence, to know the future as embodied in things and events, embedded in processes and as carrying forth the deeds of the past. Our contemporary situation entails that we understand ourselves not as objective observers and voyeurs but as implicated participants, inescapably responsible for that future in the making*, irrespective of how far down the line latent* effects may emerge as symptoms.

From the perspective of a re-embedded processual future we see the world anew. We re-cognise, that to trade and discount something which does not belong to us is theft after all. The futurity of matter and the aspirations of others as well as future peoples' needs and rights begin to re-emerge from the shadows. All that is air congeals into form, becomes tangible and real. We can take responsibility for our dreams and aspirations projected into products and processes. We can accompany latent, immanent, interconnected process-worlds of our making to their realizations sometime, somewhere.

CHAPTER TWO

THE FUTURE TOLD

Introduction

"Knowledge of the future is a contradiction in terms", writes Bertrand de Jouvenel (1967: 5). Yet, despite this contradiction, futures have been told since time immemorial and forecasting the future is something we still do on a daily basis. All of us are prophets, predictors, prospectors and planners of the future when we negotiate traffic, keep appointments, honour obligations and commitments. The future is envisaged and assumed when we explain what we will be doing today, tomorrow and in the more distant future, when we declare that we are going on holiday in three months' time, that we are learning to drive a car and taking out insurance for it, that we are moving house, changing career and signing an employment contract. All these projections and plans imply knowledge before the event and depend on a substantial stock of experience and tacit know-how. In our daily lives we move in and out of such different futures without giving much thought to the matter, treating many aspects of the 'not yet' as known, rarely attending to what it is we do in such situations and how we go about doing it. When the personal reservoir of knowledge appears insufficient, there has been and still is a tendency to turn to experts who have specialised in particular aspects of telling and foreseeing the future.

In this chapter we focus on the knowledge element of our constellation of action, knowledge and ethics. We consider who have been and who are still thought of as experts on the future, examine the sources of their specialist knowledge and survey the methods employed. We show that it is of practical significance whether the future is conceived as pre-given and actual, as empty* possibility or as virtual realm of latent* futures in the making*. We indicate that ownership of the future has knock-on effects for the way the future is perceived and responsibility anchored. If the future belongs to god(s), for example, efforts to know it are more likely to involve discovery, disclosure and interpretation of destiny, fate and fortune. If it is tied to the cosmos then calculation, prediction and extrapolation of planetary movements and auspicious moments for change may be involved. If, in contrast, the future is seen as ours for the making and taking then imagination may be employed for conjecture, creation, colonization and control. Utopias* may be constructed and pursued. Once people's relationship to the future changes from fated recipient to that of protagonist and agent of change (Peccei 1982: 11) the locus of responsibility changes too. It is moved from its external position to the new owners and protagonists. The onus is on them to know their projections and productions, including associated potential ramifications, in order that they may accompany these creations to their eventual outcomes.

The chapter takes the reader on a journey that extends from early Western cultural activity to the contemporary world of planning and producing futures by scientific, technological, economic and political means. Along the way it considers the many varied tools that have been employed to know the unknowable, to achieve glimpses of the not yet, gain knowledge before the event, provide advance warning, conjecture about possibilities and prepare for uncertainties. It familiarises the reader with practices of divination*, prophecy*, prediction, forecasting, foresight* and scenario planning* to offer comparative analyses that establish both continuity and distinctions between futures told across the ages.

Glimpses of Fate and Fortune

To divine the future is to engage with a *future present**. It is to expect a future that can be known, 'seen' and anticipated. Unlike, for example, the future of contemporary scenario planning which is open* and defined by potential, the divined future tends to be pre-given, ready set out with little room for manoeuvre or influence. Divination therefore is an effort to know what gods and fate have in store for individuals and collectives. Furthermore, it is not the 'future in general' that is being sought but answers to specific questions about what will happen in a certain situation or to a particular person.

In ancient civilisations diviners were experts that tended to be held in high social regard. They advised sovereigns on all aspects of their rule, providing guidance for both mundane and life-changing decisions. From archaeological finds we know that their craft was taught and handed down through the generations. Thus for example, *hepatomancy*, the inspection and interpretation of the surface and cavities of the liver, was practiced in the service of sovereigns. It was used to foretell impending disasters and as guide to potential actions. Archaeologists have found material evidence of this practice dating back some 4000 years to ancient Mediterranean civilisations (Lewinsohn 1961: 55). Caesar's demise, for example, was foretold by this method and prompted his appointed psychic, Spurinna Vestricius, to counsel Caesar, "Beware of the Ides of March" (Shaw 1997: 99).

In her encyclopaedia of divination Eva Shaw lists some 1000 entries of both ancient and modern practices and practitioners of divination. Many of these specialist activities end in the suffix *-mancy*, which is derived from the Greek *manteia*, meaning divination and has its root in turn in *manteuesthia*, to predict and *mantis*, the prophet (Shaw 1997: viii).[1] Thus, for example, *aeromancy* is the interpretation of cloud and wind patterns, *cleromancy* the reading of bones and other shaped objects that are thrown, while *geomancy* draws inferences about future happenings from the patterns and shapes of natural objects. The suffix identifies the origin of the practices in Mediterranean cultures. Other divinatory traditions extending back as far and further into pre-history are associated with Western cultures extending from Northern Italy and Germany to Iceland and Ireland, which are known to have consulted runes since the Neolithic period. The runes are marked with symbols and are cast much like modern dice (King 2000; Shaw 1997). Divinatory activities of Celtic cultures, in contrast, are more difficult to identify since these oral cultures have left neither written records nor artefacts that assisted their sages' extension into the future present. Their divinatory practices are preserved almost exclusively in mythical stories and song where they are associated with great powers of vision and foresight (Wood 2000, Zohar 1983).

Archaeological and historical evidence suggests that there has been an immense diversity of divinatory practices, and yet we can also discern some unifying features. What these traditions share in common is an assumption, first, that there is a *pre-given future present* to be known, and secondly, that it is our vision, our capacity to see this future which is imperfect, that is, clouded and shrouded in some way. With sufficient practice and perseverance, it was therefore thought, we may be able to read the signs, interpret the patterns and gain a clearer vision of what

[1] It is for this reason that Shaw prefers the term *hepatomancy* whereas Lewinsohn uses the term *hepatoscopy* to refer to the same practice—both terms are in general use.

nature, the cosmos and god(s) have in store. Thus, divinatory practices afford chosen specialists access to this opaque realm beyond everyday reach. These specialists, in turn, aid people's efforts to be prepared for what is to come. Their assumptions are both similar to and different from approaches to the future that are based on reading planetary patterns and establishing connections to the future encoded in nature's processes.

It is this distinction we want to explore next by differentiating between shamanic and astrological ways of telling the future. Shamans and astrologers share a common goal. Both seek to connect the human social sphere with cosmic forces, that is, to link the personal and social world with the patterns and energies of the universe. The means they use to achieve these ends, however, differ significantly and so do their respective underlying belief systems.

The shaman is an ancient figure whose magic was (and to a lesser extent still is) valued in cultures across the globe: East and West, North and South (Drury 2000; Lippincott et al. 1999; Shaw 1997: 236–7). Shamans act as bridges between the terrestrial and celestial worlds, between earth, gods and spirits. For Australian Aboriginal shamans, for example, the extraterrestrial world is the dreaming time, the realm of creation and destiny where everything is prefigured and to which all souls return at the point of death. In the most general sense and, irrespective of specific cultural traditions, shamanic practice is concerned with the wellbeing of souls in a universe where everything is imbued with a soul: animals, plants, rocks and mountains. Thus, for example, much reparation work has to be done with respect to the souls of beings that are consumed to ensure that the spirit world is kept in balance. Writing about shamanism in the arctic region, Nevill Drury quotes an Iglulik shaman who acknowledges this very problem.

> The greatest peril of life lies in the fact that human food consists entirely of souls. All creatures that we have to kill and eat, all those that we have to strike down to make clothes for ourselves, have souls, souls that do not perish with the body and which must therefore be pacified lest they should revenge themselves on us for taking away their bodies (Drury 2000: 16).

In the course of their important work shamans are able to leave their bodies and journey to the sky, the depths of the seas and beneath the surface of the earth where they make contact with the spirit world, seeking atonement or asking for guidance, advice and help. They are able not only to transcend the spatial limits of earthly existence but also the temporal boundaries imposed on terrestrial life: they move with ease between past, present and future, from whence they report back to the present. The shaman, we need to appreciate, is not a medium but an *intermediary* and a *mediator*. Shamanism is active, seeking out the spirits and souls to be consulted and concerned to keep cosmic energies in balance. Shamans are chosen ones whose power is both earned and bestowed, involving not just extensive personal development but initiation rites that take them to the realm of death where significant parts of their being are exchanged for ones that aid their visionary activities. Despite the altered state of consciousness, which is an integral part of their specialist practice, however, shamans take full responsibility for each of their journeys and the respective outcomes.

To appreciate the difference between shamanism and other divinatory practices, it is helpful to briefly consider *necromancy*, the communication with spirits and souls of the dead, associated with the ancient world of the Middle East, Greek antiquity and the Old Testament. In this divinatory tradition the dead were thought to have privileged access to the future but since to wake and unsettle them was considered a dangerous enterprise, specialists were needed to conduct the ritual investigations. Babylonians, for example, had special priests who were experts in necromancy (Lewinsohn 1961: 65). However, while their subject matter and their role as mediator often coincided with that of the shaman, necromancers were neither expected to enter those realms themselves nor did they tend to be held responsible for the pronouncements resulting from their mediations.

The ancient civilisations of Assyria, Babylonia, Egypt, Persia, Greece, India and China as well as those associated with the empires of the Mayans, Incas and Romans all shared in the belief that human fate is connected to the stars and the gods associated with these heavenly bodies. All are known to have relied on astrology to guide their decisions, appease their gods and ensure their collective prosperity. Like shamans, astrologers are active mediators but unlike their counterparts in divination they can practice their craft without the need to journey to the worlds they seek to connect. Their knowledge draws on accumulated recorded wisdom that links the movements of stars and planets with personal character traits and destinies. Like many other divinatory practices, astrology requires knowledge of patterns and understanding of interdependencies and, in addition, it relies on highly complex calculations. The resulting birth charts identify pre-dispositions as well as potential and auspicious moments for action.

A first change in the fortunes of this ancient knowledge system is thought to have occurred when Aristotle began to shift his allegiance from astrology to the science of astronomy. From this period onwards, there has been a noticeable decline in the public status of astrology. Christianity and later the rise of science have further dented the collective appreciation and acceptance of astrology as a means to tell the future, but they have by no means eradicated its popularity. Thus for example, as late as the 17th century in France, the chair of Royal Professor in Mathematics was occupied by the astrologer Jean-Baptiste Morin, who was renowned for his powers of prediction (de Jouvenel 1967: 49). Furthermore, astrology continued to play an important role in medicine. Since every part of the body was thought to be related to a sign of the Zodiac, the timing of operations and other medical interventions was considered crucial. As Richard Lewinsohn (1961: 81) notes, in the Middle Ages, "any doctor who failed to consult the stars before an operation was open to charges of wilful neglect". Today, most Americans and Europeans still know their birth signs and many of them read their horoscope, even if they don't take it (very) seriously. During the early 1990s, more than 10,000 astrologers were practising in the USA (Shaw 1997: 18) and many held advisory positions in business (Alexander 1992). Furthermore, if the popular press are to believed, numerous high-ranking politicians have availed themselves of their services.

For astrology, both timing and regular, predictable motion are of the essence: the accurate time and place of a person's birth and the precise constellation of the stars at a particular moment in time are related to the right and wrong time for decisions and actions. Exact timing and regular motion are thus the sources of knowledge for this particular mode of telling the future. Moreover, unlike shamanistic insights, the knowledge produced by astrologers is external to the person conducting the enquiry. It is verifiable by others who possess the same knowledge and skills. It shares this feature with scientific prediction. However, in a socio-cultural world that thrives on scientific knowledge rooted in verifiable *causal* connections, it was the difficulty of establishing interdependencies between planetary movements, the human psyche and social action that proved instrumental in the demise of astrology as a respected science. It is this insistence on publicly verifiable causal connections that today also poses major problems for all forms of prophecy. A recent prophetic movement can serve to illustrate the point.

Of Voices, Visions and Visitations

Oracles, prophets and mediums are conduits for the messages of god(s), spirits and souls from the netherworld. They are the message-bearing bodies through which supernatural beings convey their wills and intentions. Prophets are the channels through which divine purpose is revealed. The source of their visions and knowledge is thus external to them. Moreover, whether sacred or secular in nature, prophecies disclose something that is pre-existing and/or pre-designed, yet opaque for the fated recipients. Thus, for example, both Greek and Nordic mythologies are replete with stories about attempts to avert the prophesied fate that have been thwarted, and foretold futures that invariably have come to pass. Furthermore, whereas the work of the shaman is active, that of the prophetic medium is marked by a more passive receptivity. Finally, responsibility is differently apportioned to the active mediator and the passive medium.

These relations emerge with particular clarity from a recent historical study, which reports on a Welsh religious revival movement of the early twentieth century. Rhodri Hayward (1997) draws some extremely interesting conclusions from the distinct nature of the prophetic practices of this revivalist movement's key protagonist and figurehead Evan Roberts (1878–1954). A trainee minister and occasional collier, Evan Roberts had numerous visions and visitations of the Holy Spirit. His self-perception as acted upon rather than as acting linked him with earlier prophetic traditions, such as the incident recounted in Jeremiah (1:9) when Jehovah told Jeremiah *"Behold, I have put my words in thy mouth"* (Hayward 1997: 162). Evan Roberts relinquished his own autonomy and declared himself ready to do "anything and everything; anywhere and everywhere", in service of the divine (Hayward 1997: 166).

Drawing on the work of Michel Foucault, Hayward suggests that the position of the prophet who submits to an external authority is a potentially deeply subversive one that cuts across traditional (i.e. modern scientific) understanding of identity, authorship, agency and responsibility. Moreover, it puts the prophet in a liminal space beyond blame, criticism and accountability for his or her inspired pronouncements and actions (Thaite and Thornton 1997). Women and children, conventionally the most disenfranchised members of communities, were particularly empowered by this deep-seated subversiveness of the prophetic role. As Hayward notes, their

> connection to the Holy Spirit allowed them to subvert the traditional framework of parental and community control [...] In its disruption of the familiar identities and hierarchies it revealed the contingency of the world and the emptiness of earthly authority (Hayward 1997: 170–1).

Hayward shows how, in response to the spread of subversive activity beyond the control of conventional authority, scientific psychological studies were initiated to investigate the voices and visions which were by then experienced by large numbers of ordinary members of communities across Wales. These investigations suggested that believers' visions of the future were mostly fragments of a forgotten past: of Sunday school education and of a repressed collective Celtic consciousness. They further pointed to a deeply anchored Self, an enduring identity, which Evan Roberts had been at pains to eradicate. With this turn to psychology, prophecies and visions were interpreted as manifestations of individual and collective memory. As such, they were re-encoded within the frame of scientific understanding, which projects and predicts the future on the basis of a known past.

> Through the rhetoric and perspectives of the new psychology, the meaning of prophecy was reversed. It was seen no longer as a revelation of the future but was instead interpreted as the eruption of a forgotten past... Prophecy was no longer seen as a divine sign of the intervention of God, rather it was interpreted as a psychological symptom demanding the intervention of pastors and psychiatrists (Hayward 1997: 175).

Although Hayward does not make this connection, in a contemporary world organised on the principles of rational science and built on an understanding of the autonomous individual, hearing voices and the abdication of ones agency is primarily associated with mental illness and schizophrenic episodes. Similarly, the gifts of 'seeing into the future'—clairvoyance, extra-sensory perception and precognition, for example—are incomprehensible and highly problematic from a contemporary mode of understanding that is rooted in linear causality and positivism, that is, a commitment to past-based empirically verifiable explanations. The past-based scientific perspective, we need to recognise, is no longer receptive to a way of extending to the future that was everyday and mundane during earlier historical periods. No small wonder then that contemporary western societies consider cultures that still value prophetic extension into the future as unintelligible and inscrutable. From a scientific rationalist perspective, we want to argue, the distance to such traditions is difficult to bridge. From some Eastern perspectives, attuned to synchronicity and synergy rather than linear causality and rationality, in contrast, the bridge to such knowledge is more easily built and the access route to some of its ancient wisdom kept open.

To appreciate the gulf that separates contemporary western and traditional modes of extending into and telling the future we need to turn to scientific predictions, economic forecasting and political planning, before completing this chapter with some considerations about futurology* and the lack of tools for knowing the long-term futures produced by contemporary applied science and technology.

Predictions, Projections and Promises

If we understand prediction, forecasting and projection as contemporary industrial societies' ways of telling the future then we need to grasp them in both their continuity and discontinuity with predecessor modes of accessing the 'not yet'. Of traditional methods we can say that they were efforts to find answers to particular questions and gain foreknowledge of specific fates. The associated specialists were thus concerned to reveal *future presents* and sought to access these by connecting disparate realms of existence: cosmos, nature and personal worlds or the domains of spirits, souls and gods with earthly paths and individual lives. Diviners, shamans, prophets and astrologers were and are concerned with future presents both near and distant, their skilled inquiries yielding secrets closed to ordinary mortals. Over and above the individual differences between various divinatory methods and sources of knowledge we can say that they share an assumption that the *future present is pre-given* and that it requires special esoteric skills to unlock its mysteries. Appropriate to the opaqueness of the domain of inquiry, traditional pronouncements on the future may be in symbolic form, in riddles and rhymes that leave ample room for context-specific interpretations. When we enter the world of scientific predictions we find that most of these principal assumptions and methods of divination have been abandoned.

A first noteworthy distinguishing feature of scientific prediction relates to the level of certainty that can be obtained by scientific methods of telling the future. For example, where scientists refer to cyclical and regularly occurring natural events, such as planetary motion, the likelihood that these will continue in the future is very high. If one has extensive past knowledge of such processes one can predict that in the same circumstances the same conjunctions will occur in the same way in the future. The past is the basis on which scientific laws are established and the ground on which it is possible to know the next eclipse of the moon or that water will freeze at zero degrees centigrade. The socio-historical and economic world clearly does not provide us with equivalent laws: the social past does not determine the social future. History is not an infallible guide to what is to come. Socially produced change, innovation and progress* mean that prediction of *social* futures by scientific means is a far more precarious affair. Despite this general difficulty, however, in certain circumstances the shift to scientific methods has significantly improved the degree of certainty with which social futures can be foretold. The way this has been achieved is of relevance to our investigation.

A common feature of the modes and methods discussed in previous sections has been the attempt to access *individual* futures and find answers to *particular* questions. During the late sixteenth and early seventeenth centuries, first attempts were made to tell the future not just for individuals but for aggregates of individuals and facts. General rates of change rather than individual or unique events became the focus of predictive attention. From church records, for example, it became apparent that death rates were reasonably constant over time, as were the average life expectancy, the annual baptisms and the marriages. Even the causes of death seemed to follow identifiable patterns. With the application of statistical calculations it was possible to project those known figures into the future and predict social patterns of this kind with surprising accuracy. This was the beginning of probability calculations*. As Richard Lewinsohn notes,

> Probability statements are merely projections of the past into the future, on the assumption that the causes—no matter whether they are known precisely or not—will remain the same and will continue to have the same effects (Lewinsohn 1961: 248).

It was also found that the larger the sample of data on which the predictions were based, the closer to the eventual outcome were the probabilistic projections. While this method of forecasting did little to tell individuals about their personal fates and fortunes, it vastly enhanced socio-political planning and policy. In later chapters we will see how this method of telling futures on the basis of known aggregates has led to the development of insurance and numerous other social institutions through which uncertainty has been rendered less threatening and the unknown tempered.

We can say, therefore, that, as distinct from other forms of telling the human future scientific prediction is largely concerned to forecast outcomes not of individual but collective actions and events. We can further say that scientific prediction is founded entirely on knowledge of past facts. If there are no past and existing aggregates of facts the future cannot be predicted scientifically. Thus the more novel the situation to be projected the less prediction will be appropriate as a tool for telling the future. The UK's BSE (Bovine Spongiform Encephalopathy) crisis during the late 1980s, where cattle were afflicted by an unknown prion disease that seemed capable of being transferred to humans, was a prime case in point. Scientists were confronted with a new disease for which they had no prior knowledge upon which to predict future deaths of animals and humans. This drama of uncertainty, lack of past knowledge and inadequacy of established tools was played out daily on television and in the newspapers, with journalists and politicians demanding projections and prognoses about the progression of this disease in order to be able to act appropriately, and scientists desperately trying to explain that this was a situation where science was unable to make predictions. *Without certainty of past facts scientists had no secure basis upon which to calculate the future.* Many years of research would be needed to accumulate and collate data that could then provide a secure base from which to make predictions (Adam 2000). Physicians who are regularly expected to make prognoses about the progression of their patients' recoveries from illness, would be in a similar predicament when confronted with an unknown disease. Thus we can summarise that scientific prediction relates to aggregates, is based on probability, and relies on causal chains from which futures are projected on the basis of a known past. This applies to knowledge about the cosmos, nature and the social realm. In cases where there are no past records, no relevant causal chains or no data, the future cannot be calculated.

When probability theory is applied in economic contexts it is again aggregate phenomena that are being calculated and projected, such as the distribution of income and expenditure. In addition to these and other key features that allow for past-based prediction, economists have noted regularly recurring cycles of crisis, recovery and growth. Since the middle of the nineteenth century, three cycles in particular have come to prominence (Lewinsohn 1961: 251–79; May 1996: 135–9). In 1860 Clément Juglar identified economic cycles of seven to nine years and suggested that these patterns of recurring crises played an essential and rejuvenating role in the economy. In the nineteen-twenties a subsequent generation of economists distinguished also shorter and longer cycles which they found to be both independent from and intersecting with the Juglar cycles. Thus, Joseph Kitchin, who studied the UK and US economies over a period of some thirty years, discovered shorter cycles of approximately 40 months, while N.D. Kondratieff recognized longer ones of 40 to 50 years.

Moreover, patterns were observed within the retail price index, mortgage rates, bank base rates and many more economic variables. These observed patterns proved strong predictors as long as no extraordinary circumstances arose. Thus, for example, the great depression of the late nineteen-twenties did not fit any of the mapped and projected cycles and consequently caught most economic forecasters by surprise. Since the economy is sensitive to socio-political events there is much that can and will interfere with even the most stable and established patterns of economic activity and thus thwart the economists' best calculations (Evans 1997). What is important to note here is that economic just like scientific predictions focus on *present futures**, that is, futures that are imagined, planned, projected, and produced *in* and *for* the present. As such they have to be distinguished from *future presents*, which constituted the subject matter of diviners.[2]

This present orientation is particularly prominent when economic futures are not just foretold but created through the *trade in futures*, a practice first recorded for rice markets in seventeenth century Japan (Boden 2000). 'Futures trading'* is the trade in projected prices of products for which there may not yet be a market. It is a commitment to buy and sell something of a pre-specified standard, at a pre-appointed time and at a future price agreed in the present. The predictions produced in futures markets, therefore, are extremely present-based whilst the activity itself operates in the realm of medium and long term futures with significant effects on social wellbeing. The detail of this particular way of engaging with the future is not important at this point as we shall have opportunity to revisit this practice in later chapters. Here we

---

[2] This distinction is developed throughout this text where it takes on increasing importance for our analysis. As far as we are aware it was first developed by Niklas Luhmann in his *The Differentiation of Society* (1982: 281) where he suggests that the present future is rooted in an utopian approach whilst the future present is technologically constituted. The two provide us with different options for behaviour: present futures for prediction and future presents for action that transforms future presents into present presents. While we acknowledge the importance of Luhmann's distinction, we take our analyses of the relation in rather different directions.

are interested merely to establish the principles upon which futures are told, the methods employed and the assumptions on which predicting and forecasting practices are based.

Both the past and the present seem to be abandoned in favour of the future when it comes to the contemporary practice of projection. Projection refers to an *intention* made public before the fact or event. It is practiced not just in industry, commerce and politics but also in science. In politics, for example, governments project or 'tell' the future with the aid of manifestos, promises and blueprints for action. As distinct from forecasts, predictions and futures trading, which depend on accurate knowledge of the past extended into the future, projections are pronouncements of promised futures which are planned to be produced and actualized. The source of knowledge for such projections is the actively chosen future for which nevertheless the outcome is by no means assured. To foretell the future of your own design is clearly such a dramatically different approach to foresight and prediction that one might argue it does not belong in this chapter at all. Yet we would like to suggest that to understand the difference to other forms of divination and prediction is a precondition to appreciating some of the proposals put forward by contemporary experts on the future for responsible future-creating activities. A first thing to note is that projecting the future to your own design appears to make forecasting substantially less hazardous than when employing the other methods we have outlined in this chapter. On closer inspection, however, it becomes apparent that it is by no means easy to fulfil the pronounced promises and to bring such projected intended futures to actual fruition. How the promised future is approached, it seems, makes a crucial difference to whether or not projections can be actualised.

In a later chapter (pp. 132–134) we outline the difference between understanding the future as an architect who works to a blueprint and an artist such as a sculptor who allows the material to play a key role in the shaping of the object. The former follows a pre-defined critical path from inception to conclusion while the latter allows for surprises and diversions, takes account of interdependencies and is sensitive to the unique circumstances that impact on and thus play a role in shaping the end result. Successful governments, that is, ones that fulfil their pledges and achieve their projected futures, seem to relate to their manifestos and election promises more like sculptors than architects. Their aspirations and plans for their country may be thwarted at many a turn, not least because they are globally tied into networks of economic, political, legal, military, environmental and many other relations. This interdependency means that the known future of their promise is not theirs alone to give and fulfil, that literally millions of others partake in the shaping of that promised future, thus have a hand in its coming to fruition or being thwarted. The promised future may be 'known' but in the case of manifestoes, for example, it is the attempt to fulfil the promise that is full of pitfalls, fraught with danger and characterised by uncertainty. Thus, to know the projected pre-set future is not yet a guarantee of knowing the outcome. It is worth noting therefore that in this particular case efforts to tell the contemporary future are beset by difficulties that would have been familiar to diviners of all ages.

On first sight, promises and projections of future outcomes of present activities are not scientific means of engaging with the future since scientific predictions are established on the basis of a known past. On closer inspection, however, we find that projections emanate regularly from the laboratories of medical research centres, pharmaceutical companies and many more institutions where science finds application. Thus, for example, with nuclear power the public was promised electricity 'too cheap to meter'. With geno-technology it was the prospect of cheap, nutritious food to feed the starving, wholesale modernisation of agriculture and cures for numerous genetic diseases that were to sway doubters and sceptics. The promises, projections and visions of potential issuing from the various branches of science, we need to appreciate further, are no more certain of their predicted outcome than those made by economists or politicians. They are subject to the same delimitations and thus just as vulnerable to disappointment. Here as everywhere else, certain conditions and interdependencies influence the projected outcome: the more innovative the practice, the less secure is the basis from which to make accurate projections. Equally, the more socially interconnected the activity, the more chance there is for interference and derailment of the plans. Both these conditions have inevitable knock-on effects for the fulfilment of promises, irrespective of whether those making the promises are economists, politicians or scientists. In the case of projections, therefore, scientists have no privileged position with respect to the certainties of their promised results because they have abandoned the territory of secure past-based knowledge upon which the logic of their investigation of the future is founded. Contemporary experts on the future have set themselves the task to engage with these issues and confront the difficulties associated with knowing and telling the uncertain future which we have begun to outline in this chapter.

Their work and some of their central insights therefore will be our focus in the next section.

Possible, Probable, Preferred and Produced Futures

Today's experts on the future—futurologists, futurists and foresight specialists—tackle the problem of uncertainty head-on and acknowledge that futures studies are necessarily concerned with a combination of *possible, probable* and *preferred futures* (Bell 2003/1999). This means that unlike their traditional counterparts, these experts have abandoned expectations about a pre-existing future, and assume instead an open future that is yet to be imagined, designed and produced.

To set out the parameters of the task, most contemporary experts on the future refer to a distinction which Betrand de Jouvenel established on the first page of his path-breaking *The Art of Conjecture.* De Jouvenel (1967: 3) directs us to the Latin terms *facta* and *futura**. The former, he explains, refers to past events, done, achieved and completed, the latter to that which has not yet come about, something that will become a *factum* only after it has occurred. While the one has already taken (unalterable) form the other is still open to influence and thus "capable of ending or being completed in various ways". Brumbaugh (1966: 649, cited in Bell and Mau 1971: 9) refers to the distinction in a slightly different way when he asserts that, "there are no past possibilities and there are no future facts". In both cases the past is closed to influence, thus open to factual knowledge while the future is open to choice and efforts to colonize and control, and thus closed to factual inquiry. It is for this reason that de Jouvenel prefers 'conjecture' to any of the other concepts open to us when talking about and seeking to know the future. Acknowledgement of the conjectural nature of any extension into the future, he suggests, helps to avoid illusions about a knowable future and confusion between *facta* and *futura.* De Jouvenel's dichotomy stands like a motto above most serious futurist writing. The distinction, and the particular quality of the future that arises from it, is taken for granted, has assumed the form of 'naturalised fact'. To date, most methods and approaches of futures studies flow from its foundations. When the temporal realm is divided into *facta* and *futura,* we need to appreciate, then past efforts to tell and know 'the' future have to be considered futile since, on the basis of that distinction, 'the' future does not pre-exist but is instead open, yet to be formed, shaped and designed. Thus, from this perspective, not 'the future' but present possibilities for the future are real. Not *future presents* but only *present futures* are amenable to contemporary futurist inquiry.

In order to distinguish knowledge of socially constituted open futures from scientific predictions of natural processes, Waskow (1969) coined the term 'possidiction', by which he means the search for *real possibilities*, which is dependent on the social investigator's skill in identifying desirable seeds of change that might flourish given the right socio-economic and political conditions and actions (Bell and Mau 1971: 36–7). 'Possidiction' therefore entails examination of the 'actually possible' rather than of past-based repetitions, which are the subject matter of scientific predictions and projections of economic trends. This approach, Wendell Bell and James Mau (1971) argue, has methodological consequences. It means that the future is imported into the present where different possibilities are actualized on the basis of images of the future. It further signifies that deterministic assumptions are inappropriate for such inquiries since the future remains open until it has become present and past. Finally, it means that the future is relative to the frame of reference employed. For futurists, therefore, the future is primarily a possible, *present future*, a future that is pictured, planned, projected, pursued, and performed in the present.

Since contemporary futures study is an engagement with *futura*, that is, with the realm that is still open to our influence, numerous tools and methods have been developed to study not just probable and possible but also preferred futures (Bell 1997/2003; Bell and Mau eds 1971; May 1996). This means the focus is not so much on past-based projections, but more on what is likely to be and more importantly still, on what could be and what should be. In the latter two cases the knowledge base is the present from which possible and preferred futures are imagined, devised, constructed, planned and pursued. To de Jouvenel (1967: 5) it is this openness which makes the future "the only field of power, for we can act only on the future". While acting on the future is not just possible but essential, however, knowledge of potential outcomes of such future-creating actions is inescapably uncertain, hence, "a contradiction in terms". To make best use of that power in contexts of inevitable uncertainty, therefore, de Jouvenel suggests that we need to understand "emerging situations" while they are still in flux and therefore subject to influence, that is, before they become facts. At the same time, however, we need to appreciate that the greater a society's capacity for innovative change, the less it can rely on the scientific source of that change to provide knowledge about the consequences of the processes thus set in motion. This led de Jouvenel (1967: 275) to state that "our knowledge of the future is inversely proportional to the rate of progress".

The implication of de Jouvenel's analysis is that in contexts of accelerating innovation, knowledge of the future is moved progressively closer to the present and knowledgeable extension into the long-term future recedes ever further out of reach. Nonetheless, the retreat to past and present-based knowledge of the not yet, that is, to *present futures* only, should not be accepted, given that our actions in contemporary society reach into ever more distant futures and cast ever longer shadows. Secondly, we need to avoid the conclusion, which arises from the distinction between *facta* and *futura*, as outlined above, that there is no pre-figured future to be known, nothing to be foretold beyond the 'factual' patterns, cycles and rates of change that continue from the past and are thus amenable to study with the past-based tools of scientific investigation and extrapolation. Instead we want to argue that knowledge of *future presents* is a precondition to engaging responsibly with futures of our making, that is, futures already set on the way whose effects extend into the very long term future, from hundreds to thousands of years. Before we explore in the last part of this chapter what might be involved in contemporary attempts to know such distant future presents, let us summarise first some of the principles of knowing the future so that we may begin to recognise what aspects of past knowledge and methodologies may be helpful for the contemporary task.

The distinctions we have sought to foreground in this chapter have been between three forms of knowledge about the future. The first has been concerned with attempts to extend into particular and unique future presents—individual, socio-cultural or natural. The second was focused on efforts to know futures that are continuities from the past based on the probability of aggregates and constellations of facts recurring. The third, finally, entailed endeavours to map possible, probable and preferable futures as bases for future-creating choices, decisions and actions. Thus, the wide variety of approaches discussed above can be differentiated according to where on the temporal spectrum their primary source of information is anchored, that is, in the future, past or present. Notions that we can foresee, presage, envisage, predict or prophesy the future, as Graham May (1996: 113) points out, suggest knowledge before the event but do not entail influence over the future. Thus, divination in all its various guises implies that it is possible to receive advance knowledge and warning about events we have no influence over, in other words, that one can prepare for the future but not change it. These ideas have faded out of use as increasingly the future has become the domain not of gods but of human action, influence and power. Furthermore, when people cease to be mere recipients of others' design and move instead into the position of protagonists, agents of change and producers of fate and fortune then not just the ownership of the future but also the locus of responsibility has altered. And with that change the future has been transformed into a contingent sphere not only of human potential, opportunity, and influence but of obligation and responsibility as well. The once privileged esoteric access to futures becomes an individual, collective and public duty. Finally, the nature of the future has changed with the shift to modern, post-Enlightenment methods of telling the future. The embedded and embodied, contextually specific and uniquely occupied future has been supplanted by an open realm that is yet to be created, colonised and commodified. The task for contemporary experts on the future therefore is not about knowing that future but rather about aiding individual and social endeavours to choose wisely from a spectrum of options and preferences with their associated potential effects. While members of contemporary societies are in desperate need of such aid and assistance, the situation is far more complex and intractable than even the most sophisticated of these analyses and methods allow for. It is this complexity we want to open up for consideration here.

With science, as we shall see in later chapters, large spheres of social uncertainty have been tamed and their threats contained. On the basis of more secure footings, the pursuit of progress became not just a possibility but an almost compulsive endeavour. As Max Weber (1969/1919) suggests in his seminal essay "Science as a Vocation", the relentless pursuit of progress became part and parcel of the more general and all-embracing rationalisation of social life with its paradoxical and often irreconcilable outcomes, and the associated tendency to undermine the original intent behind them. In this essay Weber (1969/1919: 137–8) argues that science is chained to progress and the competitive search for the next innovation. Importantly, he shows how this quest for progress has the paradoxical effect of *reducing* rather than increasing the certainties and securities that are being sought.

Today this diminished stability is evident and felt in all spheres of social life. Trusted structures are disintegrating, reliable continuity evaporating: the job for life, the company for the next generation, the security of energy supplies. The uncertainty that accompanies social change has emerged as one of today's most reliable certainties. As this uncertainty increases, furthermore, our knowledge of the future is being continuously foreshortened, compressed and reduced to the present while the effects of our activities extend ever further into the distant beyond. As latent processes* these time-space distantiated[3]* implications of past and present practices are real if not visible and material in the conventional sense. As virtualities they are affecting all they come in contact and connect with on earth, in our stratosphere and the cosmos. Not just 'big science' and considerations about, for example, where to dump our nuclear waste, but the humble fridge, the trip to the supermarket, the holiday on distant shores, all contribute to the expanding gulf between our ever decreasing knowledge of the futures of our making and the vastly increasing time scale of our actions' effects. Refuge in 'futures in the mind' is clearly not an appropriate solution, although recourse to imagination is.

The 'realness' of our futures in the making, we want to stress, however, is different from the one alluded to when Wendell Bell (2003/1997: 76) suggests that "present possibilities of the future are real". Bell has in mind 'dispositionals' usually recognizable by the suffix 'able', 'ible' and 'uble', which tend to describe possibilities. One example would be that glass is breakable if dropped. The *condition* of its breakability is real irrespective of whether or not the glass is dropped. Thus, for a glass goblet breakability is an unrealized and dormant "real present possibility for its future". Process futures of our making which constitute latent future presents, unlike the unrealized disposition of the glass goblet, are already under way. It is therefore no longer a question of choice (or accident) whether or not the real disposition is going to be actualized.

To understand the distinctiveness of futures in the making, and to appreciate why neither prediction nor engagement with probable, possible and preferable futures are sufficient to know and engage responsibly with these socially produced futures, we first need to return to the distinction between *facta* and *futura*. This distinction is a crude and static simplification whose specific framing of the issues brackets and thus bypasses the temporal complexity of the contemporary condition on a number of counts.

[3] Anthony Giddens' (1984) widely used term for the stretching of phenomena across time and space.

First, it misses the 'factuality' of past futures that are in progress, futures already under way in our present, set in motion but not visible because they have not yet materialised into empirically accessible phenomena. Prominent examples would be the long-term effects of radiation, chemical pollution and global warming, that is, of processes already in progress that have not yet materialised into *facta* in the conventional sense. Clearly, such processes are neither mere aspects of our imagination nor territories fully open to our influence, power and desire. They may be knowable as partial past facts and projectable continuity from the past. Their existence may constitute, for those seeking to know, an assemblage of near certainties, contingencies, constraints, virtualities, uncertainties, indeterminacies and 'unknowables'. Furthermore, they may be recognisable to our bodies at the cellular level, our cells having absorbed and incorporated some of these empirically inaccessible features of reality, setting in motion invisible processes that will emerge as cancers, hormonal disorders and evolutionary mutations sometime, somewhere.

Secondly, the distinction creates an illusion of the future as an empty vessel to be filled or an open territory to be occupied and colonised. The distinctiveness of this emptied-out future will concern us extensively in the chapters that follow where we will show that the contemporary future is always already occupied with the latent outcomes of choices, desires, decisions and actions of predecessors and contemporaries. Unlike the pre-set future which was the subject matter of divination the contemporary future of prediction, projection, planning and production is a 'territory' in which each generation is both trespasser *and* temporary tenant as well as a temporal realm where past, present and future materiality, processes, patterns and knowledge practices* intermingle and interpenetrate.

Thirdly, it fails to differentiate between efforts to know *future presents* and *present futures*. Both pertain to *futura* rather than *facta*. The difference, however, is essential if we want to grasp what distinguishes forms of divination from prediction and futurological approaches. Thus far, the scientific mode of inquiry has no tools with which to engage with future presents. Future studies, in contrast, might have appropriate tools, such as scenario planning, horizon scanning or back-casting, all of which place the investigator's object of inquiry in the future present.[4]

[4] See for example Bell (2003/1997), Bell & Mau (1971), Inayatullah (2005), May (1996), and Slaughter (1995).

From that vantage point, in turn, futurists seek to devise the means and paths to reach desired end states. While the importance of these methods must not be understated, it needs to be acknowledged that those future presents, with few exceptions such as Bell's (2003/1997) dispositionals, are not conceived as *facta* but as products of the imagination only. That is to say, with few exceptions, the future presents of future studies are predominantly conceived as aspects of mind. They do not have the reality status of ongoing processes in a state of latency, as outlined in the first point above. Thus far, therefore, it seems as if only prophets, diviners, shamans and astrologers had sought the tools and means to connect with non-materialised yet real, specific, embedded and/or embodied future presents. Although we are not suggesting that we should therefore resurrect any or all of the divinatory practices, we want to argue that to re-visit those traditions would be advantageous, since it would allow us to explore which, if any, features of the ancient knowledge practices might be helpful for the contemporary endeavour to know and foretell latent future presents.

From the above it becomes clear that today a fourth future requires our urgent attention. In addition to probable, possible and preferable futures there is a need to know the *produced* latent futures of our making. This, as we have indicated, requires different methodological tools, different assumptions and a different temporal base for our knowledge. It entails that we re-engage with *future presents*. Much of this book is an attempt to come to terms with this new task which arises from the contemporary condition and for which we have no appropriate past or present models to guide our efforts.

Reflections

Ownership of the future, as we have seen, is intimately tied to what can be known about this temporal realm and who are considered experts on it. When fate and fortune were in the laps of gods the task was to find access to a pre-existing realm through various forms of divination and prophecy. When in contrast the future is a human creation, it is transformed into an empty 'not yet' that is subject to our will and design. The respective expertise differs fundamentally between diviners who seek to access future presents and contemporary scientists, economists and policy makers who produce present futures and attempt to know potential outcomes. Yet, the either-or choice between the two approaches breaks down in the contemporary context where future making creates latent future presents that stretch across unbounded time and space. Such futures, created in the past and present, are in progress. They are neither pre-existing in the earlier sense of fate, nor purely an aspect of our mind, will and desire. To know contemporary futures in the making therefore requires expertise that is not fully encompassed by either divination or conventional past-based predictions. In this book we begin to map this new futures territory, unravel some of the processes involved and give serious attention to the inherent interrelations between action, knowledge and ethics.

CHAPTER THREE

THE FUTURE TAMED

Introduction

In the previous chapter we focused on how we know the future. In this chapter we shift emphasis to know-how. All knowledge, we need to appreciate, is tied to action, hence our use of the concept 'knowledge practice'*. However, if 'how we know' is to be found on the knowledge end of the spectrum of knowledge practices associated with the future, 'know-how' occupies the practice end. It concerns practical knowledge which tends not to be reflected upon and theorised about. It is knowing what to do and how to go on without necessarily being able to provide a worked-out explanation. In The Future Told we explored the most familiar futures territory which has been extensively studied and utilised in contemporary practices across the full range of public social domains. While crucially important to any wider understanding of the social relations of the future, how the future is known nevertheless forms only a very small component of those relations and approaches. In this and the next chapter we therefore investigate the less familiar terrain of know-how as expressed in response to some existential conditions of uncertainty created by change, mortality, human freedom and economic exchange.

To better understand our relationship to the future, we argue, requires that we delve back into pre-history and the beginnings of cultural existence since it is here that our earliest responses to what lies beyond experience have been formed. It is here that first attempts to tame the future have been recorded in myth and ritual, in sacred and profane activities. This chapter identifies some of these practices, considers their underlying assumptions and makes comparisons in order to better understand contemporary dilemmas that arise with an immensely increased capacity to create futures that is not matched by an equal ability to know outcomes.

The inevitability of change, be it of a cyclical or cumulative kind, has fostered an array of cultural practices in response to the problem of transience, uncertainty and indeterminacy. Change, which makes the future unknowable, is not only due to the will of god(s) and the creativity of nature but, importantly, as we explain in the first section of this chapter, is also fundamentally rooted in human action and socio-cultural existence. As embodied beings, moreover, humans are bounded by the cycles of life and death, growth and decay. In the wider scheme of nature and the cosmos, their individual lives are but a brief flicker of existence. They are of this earth, but through their reflective knowledge and freedom of action are also set apart from nature, other creatures and their earthbound existence. The existential challenges which arise with the inevitable uncertainty that accompanies this distancing are considered in the second section of the chapter and are followed by a brief investigation of approaches to mortality. Responses to change and transience, uncertainty and the inevitability of finitude rooted in mortality are therefore the focus here.

In all these challenges the path to transcendence has been one of knowledge. Knowledge, however, has not been a straight-forward blessing. In moving humans beyond their earthbound niche, it has often brought them dangerously close to the realms of their god(s). Ancient myths are replete with stories about this dual effect of knowledge: Prometheus having to endure the wrath of the gods for stealing fire to give to humans or Adam and Eve being banished from paradise after eating from the forbidden tree of knowledge are just two prominent examples. Again and again, ancient mythologies suggest that knowledge has changed the position of humans vis-à-vis their god(s) and nature.[1] With this shift in position came changes in social practice that are of central concern to us here as they help us understand the way the future has been tamed and pressed into human service. These are addressed in the last section of this chapter where we build bridges to modernity by considering the continuities and distinctions between faith in providence* and belief in progress*.

Change and transience, uncertainty and finitude each pose different problems for relations to the future. Nonetheless, these problems share one central feature: they all make it difficult to know what comes next and by implication how to act, how to go on, how to prepare and plan. Each one thus constitutes an existential challenge for knowledge practice. The first of these to be discussed relates to cycles of change.

---

[1] For myths of life, death, knowledge and the beyond see Adam (2004: ch. 1), Ferguson (2000) and Littleton (ed.) (2002).

Stabilizing Change

Every repeating cycle, no matter how similar in its return, contains within it the seeds of change. Even the most repetitive action entails asymmetry and direction both within it and in relation to its environment. No cycle, we can safely say, is ever exactly the same in its repetition. On the basis of ensuing differences we can distinguish between before and after. In an earlier work Adam (1990: 168) noted that even in the physical world the likelihood of just "one cubic centimetre of air to return in exactly the same composition is calculated as ten to the power of ten trillion years; a mathematical expression for 'as good as never'". There is no sameness in nature, only similarity. For there to be life there has to be difference, be this in the form of change, transience and/or mortality. To put it differently: without processes that produce difference life ceases. The opacity of the future therefore has an ineradicable foundation in this principle of life. Moreover, since no repeating process returns in exactly the same way, degrees of uncertainty are endemic to each system.

Cyclical similarity in nature does however provide all living creatures with a measure of predictability. This predictability is both encoded in their genes—animals know when to move to their mating grounds, for example—and set up as a predisposition to learn which allows for responses to and anticipations of context-specific differences. The longest of these cycles are found in the cosmos and associated with the movement of planets. On earth everything and everybody is embedded in annual, seasonal, monthly, tidal and daily cycles. As living beings we are rooted in the cycles of life and death, with creatures' life spans being primarily tied to their metabolic systems. Our bodies in turn are pervaded by cycles of differing lengths which are rhythmically organised and pulse in tune with their environments. At all these levels, cycles produce overall, temporally extended discernible patterns but cannot provide certainty about individual expressions.[2] Consequently, at the level of individual actions and events unexpected 'surprises' have to be understood not as the exception but the norm.

The first cultural achievement we want to discuss therefore relates to the taming of cyclically constituted difference and change. The ingenious

[2] For further theoretical work on the nature of rhythmicity and cyclical change, see Adam (1990), especially chapter 3 and the last part of chapter 7.

way this has been achieved is through the cultural transformation of cycles into circles. Circles ensure that the repetition is marked not by difference and similarity but by sameness. Exact repetition always takes you back to the same beginning. It closes the circle. Change is kept at bay. The circle of unchanging repetition is not encoded in genes but in rituals and cultural traditions which guarantee that reality continues to be (re)created in this and no other way for ever more. Words and actions are handed down through generations, their unchanging form deemed sacred and untouchable. By this cultural means cycles of change are stabilised and for each of the ritualised circles the future becomes knowable. In this first crucial step, therefore, the inescapable unpredictability of life has been rendered predictable in its cultural form. The future has been tamed.

In his seminal *The Myth of Eternal Return* Mircea Eliade describes some of these processes and although the future does not feature explicitly in his work, we can clearly discern from it how the cultural production of circles of sameness impacts on the capacity to foresee and predict, enabling action in the secure knowledge that each ritual (re)creation of reality will be unchanging and continue to be so *ad infinitum.* Archaic societies, Eliade suggests, are indissolubly connected to and embedded in the cosmos, whose history is told in mythical stories of the beginning. This reality needs to be recreated and regenerated periodically in accordance with the original model.

> [Prototypes] are repeated because they were consecrated in the beginning ("in those days", *in illo tempore, ab origine*) by gods, ancestors and heroes [...] The gesture acquires meaning, reality, solely to the extent to which it repeats a primordial act. (Eliade 1989/1959: 4–5)

Myths are enacted in rituals through which the mythical period of the beginning is not just represented but *enpresented*, that is, actualised and made real in the present. Change, which accompanies the processes of life and profane action, is therefore suspended through ritual regeneration. As such, Eliade (1989/1959: 89), argues, "this eternal return reveals an ontology uncontaminated by time and becoming [...] The past is but a pre-figuration of the future". We are reminded here of some lines in T.S. Elliot's poem:

> Time present and time past
> Are both perhaps present in time future
> And time future contained in time past. (Eliot, 1963: 189)

When change is tamed culturally through the ritual transformation of cycles into circles, and by stringently adhering to the original model, it is possible to expect a particular future with confidence and certainty. This extraordinary cultural achievement of archaic societies provided islands of stability and certainty in the vast sea of change which constitutes the base condition of all life. The cultural response to change cycles envelops these living processes within socially produced ritual order and structure. It ensures that the future is a sacred re-enactment of the past. As such it allows for extension into the future from the secure position of a people that know what is to come. Creating actions that circle back to the beginning, gathering up all of past, present and future along the way, is thus one of the key cultural means to provide maximum security and confidence in the face of what lies beyond experience.

Against Uncertainty

Like living processes, human actions occur in webs of interdependent relations. They too set processes in motion, begin a networked chain reaction of doing and receiving, giving and taking that is ultimately boundless. Acutely aware of the complexity of human action Hannah Arendt notes

> [...] the smallest act in the most limited circumstances bears the seed of the same boundlessness, because one deed, and sometimes one word, suffices to change every constellation. (Arendt 1998/1958: 190)

The indeterminacy of human action, we need to appreciate, goes deeper and is even more far-reaching than the change processes of nature within which those actions are embedded. The difference is rooted in variably constituted freedoms related to space and time. Thus, for example, plants once germinated are usually tied to their place of growth and their life cycle is fairly tightly bounded in time. In contrast to plants, animals have freedom of movement across space and, through the capacity of learning and adaptation to context, their temporal freedom is expanded. In addition, human beings have mobility in both space and time. While their bodies are bounded by the same space and time limitations that govern the lives of animals (in parts even more so because they cannot fly unaided, their capacity to swim is inferior to all sea creatures and their ability to move cannot match the speed of big cats or horses, for example) their minds allow them to move with complete freedom into open-ended pasts and futures. *These forays into time enable them to conceive of alternatives: transform cycles into circles, build islands of permanence in the sea of change, think and plan ahead, create substitutes for wings and flippers.*

Most importantly, the enhanced space-time freedom affords us the luxury to change our minds. We can take alternative routes, not follow rules, break with tradition and not do what is expected of us. The very freedom that marks us as humans, therefore, is also inescapably tied to the increased uncertainty and indeterminacy that accompany human action. The associated need to bound and delimit what is potentially boundless and limitless deepens the cultural taming of futures. Since the web of socially networked processes of actions and reactions that ensue can be neither known nor controlled, there is a need for cultural responses of a social, political, institutional and legal kind. Before we can consider some cultural answers to the fundamental openness of social processes, however, it may be helpful to trace Arendt's tripartite theory of action as it provides important insights for the issues addressed in this and following chapters.

Arendt's analysis of human action is anchored in Greek antiquity, a cultural period that reaches back further than two thousand years. According to Arendt (1998/1958), classical Greek society divided human action into three spheres, each associated with different aspects of social being and distinguished by social standing—labour as the work of the body, work as the activity associated with the hand and political debate primarily identified with the mind. And, although this is not spelled out explicitly in Arendt's analysis, each action realm is marked by a specific relationship to the future.

At the lowest level of social esteem is the *labour* domain of reproduction. The primary focus of labour is the body and the satisfaction of its needs, that is, birthing and social nurture, care, nourishment, clothing and shelter. In Greek antiquity, this is the action world of women and slaves who labour in the temporal realm of ephemerality and transience. Nothing they do endures: the children they bear and nurture succumb to disease, grow old and die. The food they produce is eaten. Clothes are outgrown or wear out. Shelters disintegrate. At the action level of labour, therefore, the only certainty is that nothing lasts. This form of action is immersed in the change processes of life.

The middle stratum of human action, in contrast, produces permanence. This is the artisan's world of *work*. Here emphasis is placed on the hand and its capacity to produce objects that often outlast not just their creators but also the societies in which the artefacts were conceived and produced. The resulting permanence ensures continuity through which the future is rendered knowable. Work thus builds jetties of solidity and endurance on the shores of the sea of change. It redeems the transience associated with the action level of labour.

The third and highest level of human action is reserved for the activities of the mind: for thinking, talking, debating and moral concerns. It encompasses both the contemplative sphere of human life and the world of moral and political action where the rigidity that accompanies the production of permanence is opened up again and reactivated into processes*. As such it is the action sphere of human freedom, the one which is furthest removed from bodily need. The freedom that ensues from action of the mind, however, comes at a price: it makes indeterminacy and ignorance of the future a fundamental aspect of human action. People operating at this level are therefore charged to find socio-cultural solutions to this inescapable by-product of human freedom and to engage with the collective past, present and future. According to Arendt, this realm, with its production of heightened unpredictability is not redeemable by yet another higher level of action but purely through the potentialities inherent in action. The potentialities she is referring to are *promise* in response to indeterminacy and *forgiveness* to counteract the irreversibility of actions and their unintended consequences. The idea of promise as a means to delimit temporal freedom is briefly discussed below, whereas the theme of forgiveness is addressed in the final chapter of the book.

Promise is fundamentally social in orientation and although it does not eliminate the uncertainty and unreliability that arises from the networked chains of social interaction associated with human freedom, promise can counteract these effects for specific purposes and in certain circumstances and contexts. To Arendt (1998/1958: 245) the power of promise lies in the capacity to "dispose of the future as though it were the present". *By bringing the promised future into the present we are able to create reliability and predictability that would otherwise be out of reach.*

> [B]inding oneself through promises, serves to set up in the ocean of uncertainty, which the future is by definition, islands of certainty without which not even continuity, let alone durability of any kind would be possible in the relationship between men (Arendt 1998/1958: 237).

Arendt, of course is not alone in pointing to the stabilising effect of promise and contract. It constitutes a key focus of social science analysis in both classical and contemporary anthropology and sociology as well as being addressed in futurist writing. Thus, for example, in the field of futures studies* Bertrand de Jouvenel (1967) identified structural and contractual securities as scaffolds around the uncertainty of the future, the former being found in the most stable cycles of nature, the latter depending entirely on creation by socio-cultural means such as traditions, rules, norms and laws. To the socially constructed securities of promise and contract we may also add the production of stability and continuity through moral and religious activity.

An interesting example where promise, expectation, interdependency and moral obligation interleave to create stability and security of relations in contexts of potential conflict is provided by Malinowski's (1920, 1922) classic study, conducted during the early years of the twentieth century among tribal inhabitants of the Melanesian Islands of Eastern New Guinea. The practice Malinowski studied refers to an elaborate exchange of gifts that ties islanders, who might otherwise fight each other, into intricate relations of obligation, moral dependency, implicit promise and expectancy. The practice is known as the *Kula* and the gifts in question are not items of need, not even ornaments. Rather, they exist for the sole purpose of being given to trading partners who in turn are obliged to pass them on to others and return a gift of equivalence to the partner who has thus enriched him or her. Arm shells circulate in an anti-clockwise direction through the islands, necklaces in a clockwise direction.

The islanders depend for their livelihoods on extensive trade of goods because their respective islands are differently endowed with soil fertility and raw materials. The trade in the non-essential valuables of arm shells and necklaces, we need to appreciate, is conducted parallel to the trade in goods. Presented as gifts rather than items of barter these valuables set in train relations of mutual obligation and responsibility that reach back to the beginning of time when mythical ancestors inaugurated the practice and prescribed its precise form forward into the distant future. Not only does this tie trading partners in the *Kula* ring into unending networks of reciprocity, as Malinowski (1920: 98) explains, it also places them "under obligation to trade with each other, to offer protection, hospitality and assistance whenever needed." Persons trading in the *Kula* ring, moreover, carry high status in their respective societies. Their status, however, is associated not with ownership and possession of these valuables but with the *capacity for generosity*—that is, with giving and the nature of the gift. Furthermore, Malinowski

(1920: 100) notes that "the *Kula* involves the elements of trust and of a sort of commercial honour, as the equivalence between gift and counter gift cannot be strictly enforced." Items are only ever held in trust for a limited period during which the item, its journey and resting places, and the generosity of its trustees become part of the stories that bind disparate groups of island communities into large social networks of obligation. The peaceful relations of moral dependency that ensued from the *Kula* system facilitated trade rooted in the secure knowledge that the ties of obligation will safeguard livelihoods and prevent war as far into the distant future as these complex, open-ended chains of interdependency would reach.

What we can surmise so far about taming the future is that the cultural achievement of certainty, stability and permanence is extraordinarily reliable and affords foreknowledge into the distant future as long as the rules established for those purposes are meticulously adhered to. Thus, the ritual transformation of change cycles into circles works as long as no variations are admitted into the rituals. Promises only bind the future into the present, and thus allow knowledge of future presents*, as long as they can be relied upon to be kept. Similarly, contracts offer future security only insofar and as long as they are honoured. Trust established on the basis of *Kula* gift exchange alongside the trade of goods is maintained only as long as everyone adheres to the rules set down by the mythical ancestors. We can further discern that the future in question is a living future* that is always already set in motion and requires ritual action to secure its materialisation. Moreover, the foreshadowed future awaiting actualisation is embedded in tradition, embodied in myth and ritual and bound to specified context. Accordingly, *for traditional societies, taming the realm beyond experience depends on connecting the ancient past to an equally distant future and binding both into a coherent unity of cosmos, spirit world and human action.* The production of certainty here is conceived as a socio-cultural task. It needs to ensure that past futures are superimposed on future presents, or to put it differently, that *future presents are matched to past futures.*

Thus, the taming of the future discussed so far, has been variously achieved through the ritual circle, the production of permanence, promise and obligation, contract and law. However, what underlies the concern to create a measure of certainty, security, stability and permanence is the uncertainty of all uncertainties, that is, death and the great unknown that surrounds it. If we lived forever, then not knowing what tomorrow holds in store, what happens next season and/or next year, or what will ensue from a particular interaction would hardly matter and certainly would not carry the same weight or importance. Mortality and finitude, therefore, need to be recognized as the ultimate root of all anxiety about the realm beyond the present with its inescapable uncertainties.

Confronting Finitude

As humans we not just live and die but we encounter death, reflect on it, ask questions about it. How death is understood, imagined, and explained has implications for who (and what) we think we are, how we explain existence, and what we consider good, right and appropriate actions in relation to that inevitability and other future unknowns. "The idea of death, the fear of it", writes Ernest Becker (1973: ix) "haunts the human animal like nothing else; it is the mainspring of human activity". Moreover, the inescapability of death seems to have created a socio-cultural yearning for immortality. Confronted with finitude some cultures have posited an afterlife in another realm of existence, others believe in reincarnation, others still expect life to be bounded by nothingness or followed by re-absorption into the great pool of life from whence they came. In addition, the all too fleeting existence on earth brings with it a second yearning for eternity. This longing too is variably translated into beliefs and knowledge practices that impact on the way the future is conceived, related to and constructed. Confrontation with finitude, Daryl Reanney (1995: xxi) suggests further, produces two features that are (as far as we know) unique to human beings: the "fear of death" and a "longing for immortality".

> This is what it means, at root level, to be human. It is a primary characteristic of the human mind that it yearns to outlive the perishable construction of flesh and bone and water that houses it. From this yearning for forever, this aching sense of passing time, springs most of humanity's greatest achievements, in art, music, literature and science. Paradoxically, it is the very awareness that life is fleeting on the wings of time that directs human activity towards the creation of artifacts that possess the durability their creators lack, images in carved stone and marble, words written in books, beauty woven from sound, ideas captured on film (Reanney 1995: xxi).

In this part of the chapter we are concerned with the threat of finitude and the quest for immortality and consider some of the ways that have been devised to deal with this existential insecurity as well as the associated desire to overcome it.

Throughout the ages people have looked to the cosmos and nature for solace, finding there evidence of the *eternal* that was lacking in their individual lives. Dying plants produce seeds that burst into new life, blossoming until their time has come to produce seed and return to the earth to nourish the next generation of seedlings. In ancient Egypt, for example, the cosmos was understood to take part in the eternal cycle of rebirth symbolised by both sun and moon and the seasonal return of Sirius on the night sky as marker of the impending floods that bestow fertility on the land. For centuries the planets' journeys across the sky have been the source of death myths and rituals. Paramount amongst these is the story about the nightly death and rebirth of the sun god Re of ancient Egypt who journeyed daily across the sky, re-entering the sky goddess Nut's womb at dusk to be reborn every morning at dawn. This imagined re-entering of the mother's womb played an important role in the way death was imagined and depicted. Many a coffin lid was decorated with the image of the sky goddess Nut and poems inviting the deceased to re-enter her womb to be re-born into the realm of the dead in which they would continue to lead a fulfilled life as long as their earthly existence was just (Assmann 2000: 27–36). This trust in the eternal cycles of life and the cosmos has been undermined by Judeo-Christian religions and finally shattered by modern science. In the former God created the earth and will bring it to an end on judgement day. In the latter the earth and the cosmos have a history which may have started with the big bang and may end in heat death or a black hole. In the Judeo-Christian belief systems only God is eternal while science has eliminated the concept of eternity altogether, leaving non-believers to confront nothingness in solitude.[3]

In traditional cultures one way of achieving immortality has been through leading a heroic life that is worth preserving in story and legend. A life marked by bravery, courage, fortitude and great wisdom would transcend the person's mortal life, effectively granting a social form of immortality. Thus, the lives of heroes are still remembered today in mythologies of the ancient world and in the lives and deeds of saints and martyrs in religious histories. In ancient Greece, for example, the heroic life was the exclusive preserve first of gods (male and female) and then of men. Heroism afforded men the status of demigod, granting

[3] The relationship of science to eternity is not addressed here as it forms an integral part of the next chapter's discussion.

them immortality rooted in Arendt's third action level of political and moral deeds. Women meanwhile laboured exclusively in the ephemeral sphere of reproduction. It was, interestingly, a religion headed by a trinity of males which was to open up the field of heroics to women. That is to say, in the Christian history of martyrdom women emerged as equals to their male counterparts, proving no less capable than men of the immortality and everlasting fame bestowed on religious heroes. Like mythological heroes, saints stand mid-way between their God and the world of human activity, bridging mortality and immortality, sacred and earthly realms.

Most traditional societies, we need to appreciate, understand the golden age of perfection to be located in the past and social life since to have been marked by a decline of varying degrees of steepness. For them, therefore, the good and just life is one that resembles the past as closely as possible. For such societies heroism is one of the rare means by which the backward spell is broken for a brief period and individual immortality is sought with an eye to both ancestors and successors. Eliade (1989/1959: 39–48) shows on the basis of examples from across the world that this period of individually based, historical memory extends over no more than forty years and how subsequently heroic deeds become absorbed into the reservoir of mythical archetypes. Collective memory in the modern age, in contrast, seems to be focused less on heroic deeds to be recounted in legend and myth and more on products: the great inventions of science and engineering, conceptual innovations in philosophy and creations in music, literature and poetry. This means that *much of what outlives modern individuals is achieved at the middle level of Arendt's tripartite schema of action where permanence is created through the production of enduring objects*, enabling individuals (both male and female) to stay permanently associated with their products: Austen's novels, Brunel's bridges, Curie's research into radioactivity, Kant's philosophy, Mozart and Beethoven's music, Shakespeare's sonnets, Newton's physics.

Next we need to explore a very different collectively constituted way to secure continuity which is clustered around the belief that there is life after death. In order to cope with the threat of mortality and the finitude of individual existence, cultures past and present have posited a range of variations on the transcendence of death. Rebirth into the domain of gods and ancestors in ancient Egyptian societies, existence in the realm of the living dead in traditional African religions, re-absorption into the ancestral realm of dreaming in Australian Aboriginal cultures, re-incarnation through a series of life forms in Hindu religion and existence in an afterlife while awaiting final rebirth on judgement day in the Christian belief system, are just some of the prominent examples from across the world and history. Religions that promise life after death also tend to provide guidance about the path to that realm, about appropriate conduct during this life to ensure eligibility and provide knowledge about what to expect once the deceased arrive there. Some belief systems place great emphasis on the journey part of the process and since the passage from death to the afterlife is considered difficult, there has been a need to provide detailed instruction to smooth the path and alleviate anxiety about the unknown.

Two such books of instruction, the "Egyptian Book of the Dead" and the "Tibetan Book of the Dead" have survived through the ages and provide insights into the beliefs and assumptions associated with those ancient texts. The former is a guide book for mourners on how to prepare the body so that it may be acceptable to the netherworld of ancestors and gods. Other ancient Egyptian texts, as we indicated in the Introduction, describe the netherworld and offer tightly prescribed rituals of embalming that re-enact the mythical journey of their god king Osiris.[4] The Tibetan book of instruction, by contrast, offers guidance for the deceased's soul and is read to the dead person in the presence of mourners, giving information about what to expect and how to behave in each of the phases between this life and the next. Both books assure grieving survivors of a life after death and that a just life during earthly existence is the first step to secure everlasting life. Followers of these two belief systems, therefore, were left in no doubt that there is continuity in another realm and were authoritatively informed about actions that need to be taken in order to get there safely. This means that despite their very different contents, both texts provide certainty about the ultimate uncertainty and both focus on action, on things that can and have to be done in order to bring the journey to a successful conclusion. It is this practical element regarding the deceased that makes these books of the dead so different from other religious texts on the subject.

All major religions today, we can safely say, provide not just assurance about a dwelling place for the dead but also hold out a promise

[4] For the Osiris myth see Ferguson (2000: 116–19) and Littleton (ed.) (2002: ch. 1). For analyses of the Egyptian Book of the Dead and other associated texts, see Assmann (2001/1984) and Hornung (1999/1997). The Tibetan Book of the Dead was translated into English in 1957 (Samdup 1957).

for a future after death, a future that provides hope and solace for the hardships that have to be endured during earthly existence. Thus, John Mbiti (1985/1969: 99) reflects that it may well be the lack of a promised future in traditional African religions that is contributing to the large numbers of Africans converting to Christianity and Islam. As African society is increasingly coming to terms with the inescapable globalised western 'open future*', a two-dimensional religion that is exclusively focused on the past and present, he therefore concludes, is no longer appropriate to the contemporary condition.

While the major religions all carry this element of hope and future orientation, not all offer immortality for believers. Despite its essentially eschatological outlook, for example, the Judeo-Christian belief system does not offer immortality after death. Rather, immortality is a gift that the immortal God bestows on believers on the day of reckoning only.[5] Thus hope based on the promise of redemption focuses on an unknown future present* which acts as a catalyst to action. As Jürgen Moltmann argues:

> Because hope stands in contradistinction to present reality, hope may not be a passive anticipation of future blessings, but must be a ferment in our thinking, summoned to the creative transformation of our reality (Moltmann 1967: 33–4).

Promise, we need to remember, is never fully resolved in the present, it connects to the past and inevitably spills into the future. When we consider more closely the nature of promise and its location with reference to past, present and future, surprising continuities emerge between, for example, the eschatological promises of Christianity and the relations of trust, obligation, hope and implicit promise that characterised the trading relations organised around the *Kula* ring studied by Malinowski. In both cases the past features strongly in what is to be expected. Both systems are grounded in patterns and decrees set out in the long distant mythical past. In the *Kula* it is the ancestors' actions and instructions that need to be followed whilst past relations governed by the *Kula* give-and-take are the basis for what can be legitimately anticipated. In the Christian belief system, God's creation of the earth, original sin and redemption through Christ's death on the cross are the long-term past

[5] This point is persuasively argued by Hoekema (1994/1979) in his detailed analysis of the Bible and the future.

that, together with the future promise of redemption on judgement day, constitute the base for action in the present which, in turn, colours expectations about individual salvation. Here, providence, promise and hope are constitutive of and strengthened by the first redemption two thousand years ago. Thus, we can see how in both systems interdependencies are created that reach from the beginning to the end of time, while the promise of future rewards is taken on trust in the one system and on faith in the other. In both cases the future does not belong to people. Instead, ownership is externally located with ancestors and God respectively. From a temporal perspective, therefore, the great divide between the so-called past orientation of traditional societies and the eschatological perspective of Judeo-Christian belief systems is difficult to uphold. When we focus on the way promise and hope are anchored in past, present and future, therefore, continuities and similarities come to the fore and differences fade into insignificance.

The practices associated with taming the future considered so far are both familiar and strange. There is both a sense of continuity and discontinuity with current industrialised societies' future relations and practices. This difference and its implications will occupy us in the chapters that follow. Here we would merely like to begin the process of opening up the issues as this helps to reflect on what it means to tame the future and to consider in which social conditions the practice of taming the future comes to an end and something else begins. The belief in progress provides the perfect tool for that purpose.

From Providence to Progress

In his classic work, *The Idea of Progress*, John Bury (1955/1932: 22) argues that providence and progress are incompatible. A true future orientation, he suggests, is only possible when the future is no longer pre-given but arises from actions in the present. In our terms this is the difference between the providential *future present* and progress tied to the creation of *present futures**. Bury shows that the modern drive to produce innovation and change requires a different past-present-future constellation from the ones that are present in the *Kula* exchange and Judeo-Christian providence. It is therefore worth our while to consider a summary of the features of and preconditions for progress so that we may be able to better understand its implications for contemporary socio-environmental relations.

According to Bury it was not until the sixteenth century that obstacles to the pursuit of progress began to be removed and a new attitude to the future began to be developed. The move to the new perspective involved the following transformations: *the future rather than the past had to be conceived as the temporal location of the ideal state, a golden age not to be returned to but yet to be created.* The theory of knowledge had to be one of steady improvement rather than decline. The purpose would be no longer to discover divine design and laws but instead to produce happiness through the control of nature. Faith in providence and final causes were to be replaced by reason and the past-based causality of the new physical sciences respectively.

Another way to think of these changes is through a temporal perspective that maps the differences with reference to their past-present-future relations. From a futures perspective the belief in progress means the future cannot be pre-set or pre-given. Instead it has to be *empty** and *open*. As we argued in the Introduction, the future needed to be emptied of all content. This means, not the providential future present but the present future is the base from which progress can be pursued. Similarly, ends are not predetermined by external sources (gods and ancestors) but established by humans for humans from and for the present. The difference can be expressed as a shift in knowledge practice from the future present to present futures. This fundamental change from providence to progress entails further that cycles and circles are opened up and flattened out to form a line with one direction only, that is, from the past to the future. The idea of progress is thus congruent with the existentialist idea that human freedom is rooted in nothingness. As Jean-Paul Sartre (2003/1943: 462–463) insists in *Being and Nothingness*, "freedom in its foundation coincides with the nothingness which is at the heart of man" and which "forces human reality *to make itself* instead of *to be*". It means that one must not and cannot allow oneself to be determined by one's past "to perform this or that particular act" (Sartre 2003/1943: 475)

This emptying out of the future, divesting it of all precedent and pre-set content, has its parallel development in the emptying of lived time and its transformation into clock time, the abstract quantity freed from contextual difference that is applicable anywhere and anytime.[6] It

[6] For an analysis of this process and its social implications see Adam (2004), especially chapter 6.

is further replicated in the sphere of economics where, as we show in the next chapter, use value is displaced by abstract exchange value. The underlying principle of the pervasive change is to replace contextuality and embeddedness with decontextualised and disembedded* relations in order to produce a world of pure potential that is subject to human design and where anything is possible. *An unintended but inescapable consequence of this change is the rise of uncertainty and indeterminacy to previously unknown heights.* When the future is actively emptied and opened, carefully honed strategies to render the natural and social world more predictable and manageable are forfeited, abandoned for the adventure of freedom. As sole authors and owners of the future, however, we also carry the sole responsibility for the outcomes of our future creating actions. It is here, therefore, that we encounter the major paradox of the pursuit of progress and the assumption that freedom issues from an open future: we are inescapably responsible for that which we cannot know.

Reflections

Competence in futurity by socio-cultural means had been achieved to a very high degree by traditional societies who managed to tame many of their unknowns. Change processes were stabilised through the construction of ritual circles. The fear of death was combated through ritual and religious practice. Permanence was established through heroism and the production of artefacts that survived their creators. Continuity was secured through chains of social interdependence and moral obligation and by locating present activities in the wider scheme of cosmos, spirit world and social relations. Today, traditions and moral codes, laws and social rules continue to be the base for what Bertrand de Jouvenel (1967: 45) called "an offensive collectively waged on the future and designed to partly tame it". What we need to establish in the chapters that follow is whether or not these social means to structure, tame and secure the uncertainties and insecurities arising from an emptied future, owned by humans and constituted in freedom, are appropriate to the contemporary condition. And, if they are found wanting, we need to consider what our options might be and what openings for change might be available to close the gap between creating, knowing and minding our contemporary futures.

We can be assured that the need to know what is in store persists and that the need for practices to counterbalance pre-figured futures therefore remains as urgent as ever. Efforts to gain insight into providence, into the will and whims of gods, the instructions of ancestors or the rules governing the netherworld may not be the most appropriate means to deal with contemporary futures in the making*. There is a need in addition to, or instead of, these traditional practices to trace complexities and interdependencies of future creations that potentially reach to the end of time in the food chain, at the level of cells, or in the global climate. The task to tame futures remains but it has become that much more difficult when it is no longer backed up by providence and the promise of an afterlife, by prophecies and instructions, and by inviolable social networks of obligation. Today we seem to find ourselves in a situation without hope or appropriate social tools to respond to the future-taming demands that arise from the contemporary condition. Having divested the future of content and rooted human freedom in nothingness we realize that taming the future has become an altogether different social affair, one that requires extensive collective effort and ingenuity.

CHAPTER FOUR

FUTURES TRADED

Introduction

In contemporary industrialised societies, the future is represented as an empty space into which we move unhindered, its vacancy allowing us the freedom to transform and improve our lives. This understanding of the future is not just a mental image, however. It informs and drives all kinds of social practice, constituting a basic habit of mind* through which complex social activities can be coordinated. So sedimented is this assumption that it appears entirely natural. However, it has a multi-faceted history, some of which we narrate here.

The assumption that the future is empty is rooted in a set of social practices whose focus is on the *trading** of the future, and whose contingent process of development can be described in many ways. Here, we concentrate on a series of changes in social habits of mind that mark shifts in the relationship between two other ways of understanding and practically constructing the future.

The first of these is the future as *abstract**, that is, as belonging to everyone and no one. An object of scientific predictions, it is the result of extrapolating mathematical relationships between phenomena beyond the present. The second is the future as *open**, that is, as belonging to some degree to human beings themselves; as produced through human intervention supported by an awareness of freedom and potentiality. This can be usefully contrasted both with an abstract future and with a pre-given or providential future where ownership of future time lies with non-human agents. As these forms of future-orientation emerged, their relationship with each other increasingly became marked by tensions. Eventually, as we shall see, these tensions were resolved after a fashion by the emergence of a third form, the *empty* future*, in which a central problem associated with the open future, i.e. which potential future to choose, is solved using mathematical methods to quantify the prospective gains and losses entailed by each of the alternatives. With the appearance of an autonomous market economy towards the end of the 18th century, and an independent social science called economics within the same period, foundations were laid for the cultural dominance of the empty future.

The story we tell in this chapter about the historical relationship between abstract, open and empty futures will concern three cultural 'themes'. Each of these shifts the balance between the assumption that the future is a predictable result of the past, and the contrasting assumption that it is the open realm of human intervention rooted in free action.

First, *perfectibility**: in ancient Greece, neo-Platonism marked a major shift in the human relationship with the future. It insisted on the ethical duty of human beings to transform both themselves and the world around them through an intellectual quest to discover the laws which governed the universe, and to thereby know the nature of the divine. The final goal of this effort was to perfect oneself by achieving knowledge of the eternal order that underlies the transient appearance of things. This knowledge could then be used as the basis for human governance of nature, aimed at restoring to it a dormant state of perfection. These goals, together with some of neo-Platonism's assumptions about the nature of the unchanging structure of the universe later influenced the development of natural philosophy (early modern science). The goal of the natural philosopher is also to attain individual intellectual perfection through the study of nature and to transform the 'fallen' state of nature into a more perfect one. Nonetheless, there remains a crucial discontinuity between neo-Platonism and natural philosophy. For both, intellectual perfection is sought by transcending appearance and knowing eternal laws. However, natural philosophy sees nature as a mechanism without any inherent tendency towards perfectibility. Without human or divine intervention, the past, present and future of the universe would be identical in the sense of being governed by the same laws. Nature only possesses an abstract future, one which is extrapolated from knowledge of its past. As a result, God and nature are separate from each other, and so knowledge of nature does not of itself lead to knowledge of God. Humanity's role is to impose a higher order on an inherently recalcitrant nature, but it lacks an ethical framework to guide its actions.

Secondly, *progress*:* The potential for transformation yielded by natural philosophy's investigations of nature inspired early social scientists to use the assumptions of Newtonian science in their study of human societies, in the hope of discovering the eternal laws of social change and stability. By seeking to use this knowledge of the abstract future of 'social nature' to promote the advancement of society towards a higher condition of harmony, they appeared as oracles of collective progress towards a future state that had never yet existed in the world. The possibility of general human progress towards a state free from conflict and scarcity transformed the future into an *open* future, entirely permeable to collective human effort. The ethical framework which natural philosophy lacked was provided by the idea that progress had a natural destiny, determined by social laws. However, this produced a tension between two constructions of the future. If the future was truly open, then there were no constraints on where progress could go. But if this were so, then there were no natural laws to guide collective action. The early social scientists were thus faced with a dilemma. Their commitment to the mechanistic principles of science was in conflict with their belief that progress was impossible without freedom.

Thirdly, *profit*: in early social science, the study of trade and commerce as indicators of progress accompanied actual changes in economic practices, which began to liberate trade from extra-economic restrictions. The use of predictions about the future as a means of generating wealth through the borrowing of capital fed the increasing autonomy of the economy within industrialising societies. Within the emerging discipline of economics, these practices of prediction were given the support of a formal methodology based on the mechanistic model of probabilistic explanation* developed by social scientists. The development of classical and neoclassical economics made it possible to simplify the context in which choices about what futures were desirable would be made. Instead of debating endlessly about how social progress should be directed, methods of economic forecasting allowed policies to be compared quantitatively with each other in terms of their likely costs and benefits. In this way, the open future of collective action becomes an *empty* future, where the value of visions of progress is determined by their exchange value, not their use value. On this basis, economic analysis determines what should be chosen, channelling human action according to 'iron laws'. An empty future becomes a resource and a commodity*, capable of being *traded* in the present for the advantage of those alive now and their immediate descendants. Ownership of the future is assumed at the collective level for those who are living now, such that anyone can stake an individual claim. As a result the empty future becomes a fragmented future, projected, used and consumed in the myriad cases every day when a plan is formed that relies on economic forecasting, whether this is the buying of a house, the setting of interest rates, or the building of a nuclear power station that must one day be decommissioned. *Here, the capacity for transformation unleashed* *when natural philosophy separates the knowledge of nature from overarching ethical principles is taken to its furthest extreme.*

In subsequent chapters, we shall argue that a new ethical framework for action and policy is required, one which acknowledges anew that the future is, in a sense, open but not empty. In other words, it is necessary to affirm, with the social thinkers of the Enlightenment, that the future belongs to humanity, but that this sense of 'ownership' refers to human responsibility, not to any presumptive right to exploit the future for present gain. Consequently a new conception of responsibility has to replace mere obedience to the laws of 'social nature', however this is conceived.

Perfectibility Desired

To transcend transience by attaining knowledge of the eternal reality that underlies all change is the goal that links the development of modern science with the earliest currents of Western thought. Through intellectual discipline, it is held that we can ascend from the imperfect knowledge provided by the senses to the perfect knowledge provided by pure reason.

The roots of this tradition lie in neo-Platonic philosophy, which began in ancient Greece before being picked up on by early Christianity. The various strands of neo-Platonic thought viewed perfection as an individual goal achieved by emulating the divine, that is, by thinking about nature from the standpoint of eternity. Whereas, for other cultures, transcendence was associated with the kinds of esoteric study we examined in Futures Told, neo-Platonism associated it with a different kind of intellectual discipline. This was practised by investigating the physical world, in order to discover its unvarying 'essence', the fixed structures that lay behind the transient appearances accessible to us through our senses. Neo-Platonism's roots lay in Orphic religious cults, which created a revolution in ancient Greek religion. Whereas the earlier polytheistic Olympian religion had taught that the future was in the hands of the gods, whose domain should not be trespassed on by humans, Orphism taught that perfection, the emulation of the divine One, is the goal of all human beings (Passmore 2000: 43–4).

In linking past, future and present, neo-Platonism differed considerably from the practices we investigated in Futures Tamed. From Plato to Plotinus and beyond, the connection between past and future is seen as one between the start of a journey and its ultimate destination. The past, in which the world was created, comes before the present age, while the future will succeed it and see a decisive transformation of the world. This image of a journey links the story of the universe with individual lives. In both cases, there is imagined to be an inherent tendency towards perfection, which must be helped towards full realisation. The guiding light of this individual and cosmic journey is eternal truth.

In the past, the journey began in confusion—in a literal state of chaos for the cosmos, and for the individual in childhood, a state of immersion in the world of the senses. When the One creates an ordered cosmos from chaos, the universe in its current state emerges. Similarly, in growing to maturity, the individual gradually realises that the senses cannot always be relied upon, and sees the value of reason. The duty of human beings is to awaken to the potential for perfection inherent in them and in the cosmos itself. Through the disciplined use of intellectual tools like mathematics, they can understand more and more of the true order of things, and ascend to higher and higher levels of perfection. In this way, they will eventually come to understand the universe from the point of view of eternity. The knowledge that they gain in this way can aid them in drawing out the inherent potential of nature, overcoming any residual chaos and disorder, and bringing about a transition between the present age and a future one in which humanity and nature reach perfection together (Plotinus 1966/253–270: IV.3.7 and IV.8.6).

The emphasis placed on the usefulness of mathematics as a means of overcoming reliance on the senses distinguishes the neo-Platonic tradition to a degree from the rival Aristotelian one. For neo-Platonists the eternal nature of things and of the relations between them is primarily quantitative, whereas for Aristotle, quantity is only one of the ten categories through which the being of things can be understood, and is not the most important. For neo-Platonism, "the space of geometry is identical with the space of the universe" (Burtt 1959: 33). Mathematics is the key to leaving behind transience for the perfection of eternity.

To summarise, there are therefore three central assumptions in neo-Platonism that concern the way we can transcend transience. First, perfect knowledge is the ethical duty of human beings, and perfect knowledge is knowledge from the standpoint of eternity. Secondly, mathematics is the key means of attaining this knowledge. Thirdly, the knowledge thus attained helps to transform humanity and nature so as to usher in a future age of perfection. It is these three assumptions that survive, in a changed form, to influence the practices of 'natural philosophy', the early modern form of natural science.

Nonetheless, a major methodological shift occurs in the centuries separating the neo-Platonists from the natural philosophers. The neo-Platonic use of mathematics had relied on geometry, and consequently neo-Platonists had seen the image of perfection in the harmonious geometrical relationships that could be discovered within the natural world, such as the orbits of the planets. Nature was therefore thought of as possessing an inherent disposition towards geometrical harmony. The influence of different mathematical methods on the development of natural philosophy changed these assumptions. The development of algebra made it possible to look for order among the smallest constituents of natural phenomena, rather than among the larger-scale patterns that emerged out of interactions between these tiny parts (Jonas 1982/1968: 68).

Based on this new mathematics, the new methods for the investigation of nature developed in the 16th and 17th centuries utilised a bottom-up, rather than top-down, view of how things fit together. It was thought that, from the mathematical investigation of the interactions between bodies, a hierarchy of natural laws could be developed, ranging from the most general to the highly specific. In this way, explanations for the behaviour of natural phenomena could be provided in the simplified form of algebraic equations. This methodological development led in turn to a new understanding of the future.

For the neo-Platonists, all change was purposive and therefore future-directed, as it could be explained by the inherent tendency of natural processes to exhibit harmonious motion. This tendency existed because the ultimate cause of order was the perfection of the One, which all of nature, including humanity, desired. For the natural philosophers, however, all change was mechanical, based solely on a 'push from the past'. This means that, whereas the future of the universe according to neo-Platonism was ultimately a qualitatively new 'age', the future of nature according to natural philosophy was simply a series of variations in the total quantity of motion in the cosmos. This meant that, at each and every moment, every configuration within nature is produced mechanistically in accordance with the same eternal laws. Natural philosophy therefore understands the future of nature as *abstract*, given that its content is predicted mathematically from our knowledge of its past. This understanding of the future was expressed in a number of assumptions about nature that inaugurated a new, entirely mechanistic view of the world, which became known as Newtonianism.

The matter that went to make up the universe was held to be composed of what Galileo (1564–1642) called 'corpuscules' (Burtt 1959: 77–8). These hard, impenetrable atoms moved about and changed direction because of their collisions with each other. Robert Boyle (1627–1691), and after him, Newton (1642–1727), affirmed that all change is only an alteration in the quantity of motion belonging to a collection of atoms. Johannes Kepler (1571–1630) argued that the primary, or real, qualities of matter are just those that can be attributed to atoms based on their motion. Those that we seem to experience through our senses, such as colour, heat, weight, taste etc., are mere secondary effects of the real qualities of matter. This meant that all change in nature could be described in quantitative terms alone, and that the measure of change was the number of units of space traversed by a body in units of time. Time itself was thought of in terms of space, as its 4th dimension, a line divided into t-coordinates just as any point in 3-dimensional space could be given x, y, and z coordinates. Based on these assumptions, it was thought the mathematical and experimental investigation of nature would yield equations that described eternally-valid relationships between natural phenomena. In this way, the ever-changing world experienced through the senses was seen as dependent on one of unchanging mathematical relationships. Consequently, the natural philosophers stayed true in this sense to their neo-Platonic heritage.

However, this mechanisation of nature had other consequences that are vitally important for our discussion in this chapter. Whereas nature for the neo-Platonists had been an expression of the divine, the natural philosophers saw it as separate from the Christian God, a realm created by him with its own autonomous laws. Although he had originally imparted movement to the atoms from which all matter is made, once in motion they needed no further intervention from him. Nature is the realm of change that occurs within time. God himself is entirely *outside* time, existing in eternity, and intervening only to create both time and the world, and to bring both to an end at the last judgement.

From the beginning natural philosophy had a difficult relationship with Christian faith. From the point of view made possible by its use of mathematics, one can occupy the 'standpoint of eternity' without knowing God. It is not necessary to know the nature of God to know the laws of nature. This means that, whereas the neo-Platonists rooted their understanding of the perfectibility of humankind and the world in a strong ethical context, the worldview of the natural philosophers could no longer do so. The connection between natural philosophy and Christian faith was more indirect. Science was seen as a corrective endeavour, a means to free the human mind from a fixation on mere appearances. The intellectual discipline of science could help prepare an individual soul for a truer relationship with God, and for the ethical perfection attainable through faith and good works, but could not itself lead to such perfection.

Natural philosophy thus remains wedded to the intellectual and practical realm of human ingenuity, but separate from Christian salvation as such. It takes over from neo-Platonism the ideal of eternal knowledge, but loses its ethical framework. As we mentioned above, however, it is also influenced by neo-Platonism's view that knowledge must be applied practically in bringing nature to a state of higher perfection. We have seen though that, without the neo-Platonic ethical context, nature could not be viewed as having any inherent tendency of development. As a result, natural philosophers saw human control over nature as having one purpose only—to benefit humankind.

Francis Bacon (1561–1626) claimed that, although Adam had, through his sin, lost original dominion over nature, there was still possible a kind of inferior dominion, to be sought through knowledge (Butterfield 1965: 98–9). Knowledge of nature would enhance human capacities for transforming the natural world through the development and application of technology. This process of transformation was seen as potentially leading to the restoration of a kind of perfection to the natural world (Ovitt Jr. 1987: ix–x), but this perfection would be attained by making the processes of nature more attuned to human desires. Human beings could use the abstract future of nature, its inherent predictability, to their own advantage. As a result, the future of nature begins to be seen as *open* to potentially unlimited exploitation by human beings.[1]

[1] The warrant for this is not necessarily, as has often been thought to be the case, justified directly and solely by Christian doctrine itself. On the various forms of relationship with nature (including stewardship) that medieval Christian doctrine proposed, see Ovitt Jr. (1987: 70–87).

Progress Naturalised

We will now trace how assumptions about the abstract future and about the open future are combined within 18th century social science to articulate the idea that *progress* is the natural destiny of humankind. This idea was foreshadowed in the 16th century, when Francis Bacon had proposed that, with the increasing control of natural processes made possible by natural philosophy, a general advancement of human learning would also result (Zilsel 1941–42: 557). In fact, the methodology of natural philosophy effectively required a concept of continual progress. The use of experiments, and the emphasis placed by natural philosophers on the repeatability of results made the collective review of knowledge a necessity in order to consolidate it (Butterfield 1965: 101). The body of their knowledge would therefore grow incrementally over time, undergoing constant improvements.

What the concept of general progress adds to this scientific vision of an open future is an ethical framework to replace the one lost in the development of natural philosophy. This framework is shared by both 18th century social science and the early utopian* literary and political tradition that emerged in the 16th and 17th centuries. It constructs the future as a general redemption from conflict, inequality and suffering. The responsibility for creating such a future is seen as belonging to human beings in the present. A similar view of the future, as one of redemption, is also held by major religious traditions. Nonetheless there are significant differences between the ways in which religious and utopian/social-scientific understandings view this redemption. The primary cause of these differences has to do with whether the future is envisaged as the *restoration* of a condition originating in the immemorial past, or whether it is seen as the achievement, through human effort alone, of an entirely new condition.

Where the future is thought of as a restoration, ownership of the future remains in the hands of a non-human agent. This restoration is generally thought of as a transfiguration of the world as it exists, and therefore is not necessarily a literal return to a previous condition, but a 'higher' realisation of perfection, e.g. a New Jerusalem. It may depend (in various ways) on some form of ethically guided human action. The guarantee of correct guidance lies in the covenant formed between people and God. By contrast, where the future appears as the promise of an entirely new way of being, the ownership of and responsibility for the future is implicitly seen as lying entirely with humans and their capacity to understand and realise their own potential through their own efforts.

In the former case, time as a whole is a spiralling ascent to a higher state of perfection. In the latter, by contrast, time is opened out (as we indicated in Futures Tamed) and becomes a straight line along which unending progress occurs. The sociologist Karl Mannheim (1893–1947) has argued that the relationship between religious visions of future states and the past means they primarily have a compensatory function, and indeed one that tends to legitimate the role of religious authority. He suggests that an important departure from this way of thinking is when 'wish-images' of utopia, creations of human imagination that depict a perfect future brought about solely by human effort, are used to 'shatter' the apparent naturalness and solidity of the present, by suggesting that other ways of being are in fact possible (Mannheim 1936: 173). It is the acceptance of the idea of *progress* as an alternative to the concept of providence,* a kind of covenant with the human future proclaimed by secular 'prophets', rather than the divine past, that makes commonplace this shift towards the future and the accompanying refusal of the guiding role of the past. For the idea of progress to be possible, reality has to be thought of as radically transformable, independent of any divine promise.

Thomas More's *Utopia* (2003/1516) is a landmark in the development of the idea of progress, for it develops just such a 'wish-image' of the future in the context of a sociological investigation. It links the idea of a transformable social reality to an analysis of social structure, by comparing the mythical society of Utopia with contemporary Europe. Against a background of social and technological change, More provided a model for thinking about different possible societies, as contrasted with the entrenched divine order of feudal organisation (Bell 2003/1997: 9–14). Once the future appears amenable to being shaped through human effort alone in accordance with standards produced through critical reasoning about desirable directions of change, the idea of the general progress of humankind begins to emerge. In the wake of More's book, there followed a utopian literary tradition, which was complemented by emerging knowledge of the historical variety of human societies, resulting in a growing consciousness of the widely varying possible forms of human social existence. These influences meshed with the conviction that human beings were capable of radically changing their social conditions which, for example, accompanied the revolution in England and the execution of Charles I.

The sense of human possibility that grew out of these influences was reflected in the early social science tradition, which sees the future as an open one of unending progress. Despite this, the early social scientists also emphasised that there was only one morally right direction in which human societies should develop. In order to decide on this direction, a properly scientific investigation into the laws of human psychology and social change was called for. This meant that there was thought to be an inherent link between human nature and the direction of true progress. In other words, the choice of what futures to create had to be based on knowledge of the principles of human nature.

For the 17th and 18th century thinkers of progress, the responsibility for producing progress would rest primarily with an elite who possessed specialist knowledge of human nature, and who would orient the evolution of human society in the correct direction, avoiding obstacles and counter-influences. This indicated that, although these thinkers saw human history as being governed by natural laws, they thought that knowledge of these laws provided human beings with the capacity for free action. Although history was a kind of 'social nature' which could be studied on the basis of similar mechanistic assumptions to those used by the natural philosophers, its future was still open, dependent on will and decision. Although tendencies of development could be extrapolated from the study of the past and present, this knowledge freed humans to imagine new possible lines of development and act to realise them.

By the mid-18th century, thinkers such as Turgot and Condorcet had begun to view societies as natural phenomena which could be studied in the same way as any other (Goodwin and Taylor 1982: 146). Human society was drawn into the neo-Platonic space of algebraic and geometric reasoning, and thereby into Newtonian space and time. Within this new spatio-temporal context, the development of societies could be seen as a succession of states of increasing or decreasing progress, hopefully leading towards realisation of the potential for development that humans, considered as rational beings, had within them.

One of the key methodological innovations in the study of human nature involved a similar mathematical advance to that taken by the natural philosophers. A gradual algebraic simplification of social relationships according to the principles of the new science of probability made it possible to describe human behaviour in terms of general, mechanistic principles. Although individual human beings were considered capable of free choice, this choice was at the same time not random, but rational. Consequently, patterns of behaviour could be expected to recur in certain circumstances, as individuals chose to act in ways they considered reasonable. Statistical and probabilistic reasoning about the aggregate behaviour of groups of people consequently became a central tenet of the new paradigm, and aided the search for an attractive force that held all societies together, the social equivalent of Newton's universal gravitation (Goodwin and Taylor 1982: 124–5). Deducing an analogous principle from the discovery of laws of human behaviour would enable social change to be managed by those with the right kind of scientific knowledge.

Human beings were therefore seen as natural objects for scientific study. Insofar as they were rational and capable of mastering scientific methodology, however, they were also seen as agents of deliberate change. Two sets of assumptions thus governed the new social science: first, a mechanistic understanding of the cosmos made possible an analysis of society as a 'social nature' with its own abstract, predictable future rooted in natural law. On this basis, the laws of human psychology and collective association could be discovered. Secondly, the assumption that human beings were essentially rational and therefore free contributed an ethical framework for the investigation. The future was not merely abstract, but *open*: it would be given shape by a process of transformation that would realise the potential for perfection present in human beings.

But these two assumptions gave rise to a problem: which of these two ways of understanding society was the foundation for the other? Towards the end of the 18th century, the philosopher Immanuel Kant (1724–1804) remarked that in order to be able to study history at all as a domain governed by laws of progress, it was necessary to assume that "all natural capacities of a creature are destined to evolve completely to their natural end" (Kant 1963/1784: First Thesis). Kant saw this assumption as a moral requirement, and therefore a matter of faith in the destiny of humanity. It was not itself a proposition provable through the methods of scientific investigation, but gave an ethical meaning to the work of social science. There was thus an inherent tension within the habits of mind upon which the new social science was erected, between the scientific methodology that assumed the future-oriented behaviour of humans as 'social atoms' was governed by natural laws, and the ethical framework that represented humans as agents of willed and projective change.

This tension gave rise to a problem that underlay many of the 18th century intellectual debates about how to determine what the end state of human development *should* be. If the future stood open to transformation through human action, then debate concerned which of various possible routes of progress were to be preferred, and on what basis. This meant that there was a further problem to be tackled. Kant had pointed out that investigators had to make assumptions about what would constitute a desirable future for society in order to interpret its past and present. Without these assumptions, it would be impossible to understand the direction of social progress. But with them, it seemed that the mechanistic assumptions of the nascent social science would be violated: the universe, as described in social science, would once again take on an inherent direction of movement, as in neo-Platonism. Social nature would have, in addition to an abstract future, a teleological one—but the affirmation of the purpose of human social progress would rest solely on the investigator's profession of faith in a particular understanding of the developmental potential within human nature, rather than on natural laws that had been discovered through scientific investigation. This tension between the abstract future of social nature and the open future of human potential would be resolved by economics, which provided an intellectual articulation of a new form of future that was being consolidated within the societies of the 18th century: the *empty* future of economic activity.

Profit Pursued

Many of the early social theorists were interested in the growing importance of trade and commerce within 17th century Europe. The capacity of nations to increase their wealth was clear evidence, they proposed, of the natural tendency of human societies to progress. This interest led to the development of forms of intellectual enquiry in the 18th century which focused exclusively on the phenomenon of trade, just at the time when new forms of economic activity were coming to the fore, and having wide-ranging effects on the institutional make-up of societies. The processes of institutional change that support the evolution of these forms reach a peak of intensity in the Industrial Revolution, fulfilling what Karl Polanyi (1886–1964) calls the 'disembedding' of the economy from the rest of society (Polanyi 1971: 81–2).

At this point, the economy was increasingly defined by the theorists of what became known as classical economics as an autonomous realm operating according to its own natural laws. Like nature had been for the natural philosophers, the economy was seen by these theorists as especially amenable to mathematical description and simplification. Whereas social science had begun with the idea that progress was an ethical commitment, and could only be achieved through positive acts of transformation, economics opened up a different prospect. Via a thoroughgoing, statistically-based investigation of the laws of trade it was proposed that the mechanisms of wealth generation would be laid bare. As the body of economic theory developed, it was affirmed that all that would need to be done for wealth to be efficiently distributed to fulfil human needs would be to take a 'hands-off' approach to the economy. Instead of taking definite steps to advance towards a particular vision of the future, all that would need to be done would be to identify how economic processes worked, and then ensure that other social institutions did not interfere with them. Not only would this lead to progress, but thanks to the economists' concentration on the allocation of wealth, a quantitative measure of progress would be provided—the more efficient this allocation, the more progress would have been made. This meant that the tension within social science that we examined in the previous section could be eased: no difficult choice between alternative futures was necessary, as scientific investigation of the laws of economics would make the best course of action clear to all rational beings. Again, as in the worldview of natural philosophy and social science, using mathematical analysis to penetrate beneath a confusing world of appearance to a stark, crystalline, essential reality would enable the future to be understood.

The habits of mind that formed the assumptions of natural philosophy began to encompass economic phenomena with the work of Richard Cantillon (c.1680–1734), who developed a Newtonian theory of trade. Cantillon viewed the economic system as an interconnected whole of mechanically functioning parts, driven by the self-interested urge on the part of individuals to accumulate profit (Ekelund and Hébert 1997: 65). Central to Cantillon's theory, and indeed to subsequent mainstream economics up to the present day, is the idea that individuals played, in the economic system, a similar role to the atoms in the universe of natural philosophy, imparting by their activity motion to the whole. Just as the natural philosophers had used the atomic model to support their methodological commitment to mathematical simplification, the economists developed their theories of wealth on the basis of the role of individuals in production, exchange and consumption. Consequently, their conception of human future orientation proved crucial, as it served to describe the 'motion' of the individual atoms. In this way, the future as envisioned by economics was composed out of a multitude of fragments, individual futures made up of economic decisions driven by self-interest, which could be aggregated mathematically. This would allow economic predictions to be made, and, provided interference from other institutions (such as the government) decreased, would enable individual economic agents to plan their futures rationally and with confidence.

In this light, the early economists argued that the process of exchange is motivated, in the first instance, by the need to secure one's life and the pleasures that allow one to enjoy that life, but is ultimately guided by the desire to maximise one's 'betterment'. What is meant by 'betterment' in this context is the accumulation of economic power, one's capacity to control one's present and future circumstances in an essentially antagonistic social world. Writers inspired by Cantillon such as Adam Smith (c. 1723–1790) therefore saw the economic system as a general clash of individual interests, but one whose aggregate tendency leads to the promotion of the common good. Individual self-interest advances a general and continual increase in economic activity without producing huge concentrations of economic power (Smith 1975/1776: 61–2). Looked at in the round, the aggregate motion of the individual social atoms leads to a harmonious whole, and so the best future would be produced through the unfettered functioning of natural economic laws. In this way, rational self-interest takes over from the moral perspective of the social scientists as the ultimate guide for action. The overall purpose of human activity is seen as governed by the natural laws of human motivation.

The social reality this body of theory was attempting to describe was one in which economic practices had, since the late Middle Ages, increasingly organised themselves around a new understanding of the future as *empty*. The emergence and consolidation of these practices had been slowed by conflicts with the practices of other institutions, such as the Church, which, as we noted in the Introduction to this book, maintained quite different views of the future. Lending money at interest, for example, had been seen as sinful by Christians up until the medieval period, given that it involved the illicit sale of future time, which belonged to God alone. As we shall see, this conflict was eventually removed by theological innovations.

The empty future was not the same as the abstract future of a mechanistically-understood nature. An abstract future is calculable by means of extrapolation from past observances. Unlike the pre-given future that is told through divination or the providential future of religious belief it belongs to no-one as such. Rather, it describes a future of repetition, the continued functioning of an external mechanism—such as Newtonian nature or the economic system. An empty future, by contrast, is a homogeneous medium of measurement, one in which different future outcomes are made commensurable by means of something that makes them comparable, and which therefore makes a choice between them possible. Money acts as such a means of comparison between different commodities (Simmel 1982/1900: 146). It also serves as a means for comparing futures. Among the emergent economic practices, specific uses of monetary comparison meant that the *exchange-value* of different futures can be estimated. This already assumes that the future in which these outcomes will happen is within the ownership of agents in the present as a future of pure possibility, without any content.

Specifically, these innovatory practices concern the production and use of surpluses. A 'long surplus' in the form of a stock of money can be opposed to a 'short surplus' in the form of more of a perishable commodity that one needs to consume to survive. Such a long surplus represents the power to obtain any number of different commodities at any time. It thus changes one's attitude to the phenomena of uncertainty and transience which we discussed in Futures Tamed, as money capital frees its possessor from the constraints and risks that accompany the cycles of production which characterise barter and subsistence economies. For example, the possibilities increase for accumulating objects on the basis of values other than their immediate utility for survival. The satisfaction of less pressing desires, as opposed to urgent needs, becomes possible.

Importantly, owning a monetary surplus enables credit for future trading ventures to be obtained, using existing stock as collateral. This was made possible by the theological legitimation of the practices of lending money and trading for profit. The merchant's vocation and the lending of money became seen as justifiable due to the power of trade in producing wealth, and thus in serving the common good, finally being praised by the likes of St Thomas Aquinas (Le Goff 1980: 61). To lend money at interest became seen as legitimate compensation paid to the creditor for her giving up for a specific duration of future time a quantity of hoarded exchange value, and with it, a portion of her economic power.

This legitimation of credit meant that money could be lent on the basis of expectations of future returns. This enabled the production of monetary surplus for the purpose of investing in the creation of future surpluses. The future is thus no longer seen as one in which surplus production in the present gradually creates independence from the cycles of subsistence production. Instead, the future is created around new cyclical structures of production and investment, with the goal of investment being an ever-increasing return. Planning production is no longer aimed at maximising the potential for consumption, but at providing for a continual increase in the rate of profit or surplus value extracted from economic exchanges. The only measure for deciding whether the future turns out to be better than the present is thus mathematical. It is this innovation in how futures are created and traded that produces the conditions of possibility for commercial and then industrial capitalism. Practices of investment and credit change the relationship between production and consumption by constructing the future in a fashion similar to mechanistic science. The future of credit and investment is a quantitative scale used for estimating the size of a return based on different patterns of investment. This future stretches into the long term, and is empty but for these results of economic forecasting practised from the standpoint of particular interests in the present. The desirability of a particular future outcome can be measured by determining the costs and benefits of pursuing it relative to others. In this way, the value of a particular present future* is determined solely on a quantitative basis: does it lead to a larger profit than the alternatives? If so, then it is a more rational course of action. In this way, the future as such becomes tradable: one future outcome is tradable for another, on the basis of its estimated returns.

The goal of accumulation through the increase of rates of profit is ultimately to secure economic power to enable control over one's own future. Success in amassing surplus or attracting investment in the present produces a corresponding increase in power to transform the future. The goal of transformation is therefore to ceaselessly augment the power to transform. Accompanying these economic practices is therefore a final shift in the ownership of the future, away from the divine and towards humankind, which is correlated with its emptying: the only content for this future is a projected increase in the power to transform reality through further economic activity.

This new economic reality therefore sees agents working to create the future first by entirely quantifying it and secondly by using this emptied future as a resource for increasing economic power in the present. Economics supports these new practices by arguing for their rationality as ways of achieving progress. In doing so, it unites the empty future upon which individual economic transactions are premised with its own abstract future, the unfettered operation of the laws of trade. Economics takes the empty future and, using its theories of natural economic laws, applies it to the totality of economic activity in society. The processes that contribute to the maximisation of wealth are those which collectively ensure that profit extracted from economic transactions increases over time. In classical economics, it was assumed that economic laws placed natural limits on how much profit could be extracted from specific transactions. For example, it was held that certain costs made up the 'natural price' of a commodity, and that selling it at a level too far above this baseline would result in a lower level of profit overall due to other producers selling it for less. Adam Smith depicted this as a natural harmony within the market, acting mechanically like an 'invisible hand' to keep economic relationships stable. This assumption means that rationality is no longer viewed as the potential of human beings for taking individual and collective responsibility for their common, open future, but instead becomes a criterion for individual economic decisions based on estimates of relative gains of exchange value.

When classical economics shifted its methodology in the mid-19th century, creating what became known as neo-classical economics (the paradigm that remains dominant in economics departments to this day) two theoretical 'refinements' occurred. First, the use of linear methods of mathematical analysis based on differential calculus was extended and made more rigorous. Secondly, a further simplification of the psychological model of self-interest on which the 'mechanics' of economic analysis are based was carried out. According to the centrepiece of neo-classical theory, the concept of marginal utility, the fundamental motivation for human action, and thus the basic thrust of all future-orientation, is to maximize the level of gratification achieved with limited resources. This conception of human psychology was characterised by Francis Edgeworth (1845–1926) as that of a pleasure machine, for whom the passage from past to future was a clockwork transition from one unit of pleasure-intensity to another (Routh 1975: 241–2). In other words, for neo-classical economics, using the empty future as a way of making decisions, based on expected returns of profit, was a fundamental law of human nature.

The role of economics, and the empty future, as a policy tool was affirmed by the hugely influential use in France of cost-benefit analysis by Jules Dupuit (1804–1866) in the 1840s to determine the best ways to provide public goods such as roads and water (Ekelund and Hébert

1997: 269–71). As economic theory was increasingly brought together with policy and business practice, the problem of decision that had afflicted the early social scientists' conception of progress began to be eased. All that needed to be done was to conduct a rigorous economic analysis of the different proposed policies, and the option that would produce the greatest return would be immediately clear. The ethical framework of choice was replaced with a fully Newtonian framework of mechanistic analysis, within which the optimal outcome would simply appear at the far end of a calculation.

As the empty future of economics began to establish itself as the sole standard of rational planning, the fragmentation of the future that was presaged by the profit-driven practices of commercial and industrial capitalism was transformed into a social and political principle. Instead of human decisions about the future being formed within an overall value-consensus concerning human destiny, the ownership of the future had passed into the hands of individuals through mechanisms of economic choice. The only embracing context for economic decisions now became the economic 'climate' as a whole, a set of predictions about the abstract future of the economic system contained within the occasional oracular pronouncements of economists and the politicians whom they advised (Evans 1997).

The future envisaged in mainstream classical and neoclassical economics is therefore designated as colonisable by a huge multiplicity of singular economic agents (individuals, corporations, governments, etc.) on the basis of self-interest, against the background of an assumed collective claim of the present on the future. This emptying of the future also leads to its fragmentation, as the all-embracing temporal narratives of perfectibility and progress are replaced by as many quests for profit as there are agents. The essentially Newtonian assumptions of classical and neoclassical economic theory lead to the conclusion that there is a natural tendency within the economic system towards a progressive harmonisation of interests, a state of equilibrium.

Reflections

The future becomes seen as a territory which must be exploited for the sake of releasing ever-greater power to transform reality. In this way, the future envisaged by economics is the most open of futures, but the assumptions of economic theory leave no room for the ethical context within which both neo-Platonism and the early social scientists framed their understanding of the open future. Instead of being open in their sense, and therefore a field of action that requires responsibility, it is empty, and up for grabs. Economics therefore mirrors the achievements of natural philosophy: where the natural philosophers implicitly separated nature from God, seeing in it only an abstract future of repetition, economists separate the economy from the social field as such, in which the early social scientists had still seen human freedom operating. Human freedom does not direct the economy—rather, the laws of utility maximisation direct it. A continuing quest for the pristine essence beneath the confused appearance results, with economics, in the identification of human freedom with economic necessity.

Within this framework, the future is no longer seen as unpredictable in a way that requires wider cultural practices of taming. From a future embedded in natural and social processes, we have passed on to a future that is entirely disembedded* from them, and which serves instead as a means of calculation, estimation and trading. The futures of perfectibility and progress could never be traded: there was no framework within which this could be done. However, as we have seen, elements of the habits of mind that support the quest for perfectibility and progress eventually make possible the emergence of new social practices for which futures trading is the basis of all social action.

CHAPTER FIVE

FUTURES TRANSFORMED

Introduction

In this chapter we are concerned to understand modernity and the industrial way of life and seek to illuminate some of the key features of attendant approaches to the future. In particular we are interested in the deeply contradictory and paradoxical characteristics of contemporary futurity that seem to accompany transformative future-making ventures. Wherever we focus our attention, multifaceted contradictions seem to be the order of the day: efforts to control, manage and engineer the future produce unprecedented uncertainties. The insatiable appetite for economic growth and scientific progress* seems also to create, as an almost inevitable by-product, environmental degradation and pollution as well as irreparable long-term damage to the health of all reproductive life forms. As the future is progressively emptied of content and opened up to the possibilities that we might create, non-intended consequences mushroom and planned outcomes prove ever more elusive. Moreover, with increase in individual freedom rises not certainty but indeterminacy and, as Friedrich Nietzsche (1966/1882)[1] recognized over one hundred years ago, the new unbounded potential produces a paradoxical yearning for binding values. Not surprisingly therefore, key thinkers associated with early modernity were wrestling with its ambiguities and tensions. Thus, Karl Marx famously wrote:

> In our days everything seems pregnant with its contrary [...] The new-fangled sources of wealth, by some weird spell, are turned into sources of want [...] At the same pace that mankind masters nature, man seems to become enslaved to other men or to his own infamy. Even the pure light of science seems unable to shine but on the dark background of ignorance. All our inventions and progress seem to result in endowing material forces with intellectual life, and stultifying human life into a material force. (Marx 1977/1856: 338)

Today the tensions and contradictions have not disappeared; we have merely become accustomed to them. Our senses have been dulled through

[1] See also Berman (1983: Introduction).

familiarity. That is to say, the paradoxes and strains have become a taken-for-granted feature of our lives. Once they are accepted as natural, however, they become invisible, which means we are no longer able to creatively engage with their potential on the one hand and their dangers on the other. Yet within the interstices of these contradictions and oppositions, we want to suggest, lie opportunities for change. It is therefore advisable to get to know the opposing tendencies that arise from within modernity's approach to the active transformation of the future, allow ourselves to be re-sensitized and discomfited by the powers that are unleashed when social futures are engineered and transformed. This requires that we understand their roots and grasp their potential for effecting change at a deep structural level. In this chapter we want to begin this process by exploring the promethean power* of producing futures in a variety of contemporary guises.

First, however, it may be helpful to recap on some of the major shifts in approaches to the future, discussed in earlier chapters, all of which were pre-conditions to the modern assumption that the future is in our hands and subject to human transformation. These alterations involved a change not only in ownership of the future from god(s) to people but also in assumptions about who counted as legitimate experts on this domain of knowledge. As we showed in previous chapters, traditional societies transformed nature's cycles of return into ritual circles that recreated the social world in predictable form. This social achievement and its attendant advantages were forfeited when ritual circles were opened out into the linear shape of future-directed progress. With this change in approach, ancient chains of obligation and promise that stretched from the beginning to the end of time had been broken and were substituted with an abstract money economy and discontinuous relations guided by market utility. In this chapter we will attend to such severed connections and focus on the paradoxes that arise when embodied and socially embedded people operate in abstract, dis-embedded and de-contextualised social and institutional structures. In previous chapters we also showed how the pre-given future gave way to the idea of perfectibility* and progress. Consequently, the golden age was no longer thought behind but in front of us: it became both goal and aspiration of the pursuit of progress. Together with wealth creation and what we call the 'frontier spirit',*[2] this future potential is the impetus

[2] Adapted from Jeremy Rifkin's (1994) term 'frontier mentality'.

that moves us to innovate and invent, control and colonise, transform and traverse. Here too however, it is the process of temporal dis-embedding and de-contextualisation that opens up the chasms between intent and consequences.

The shifts, mutations, changes and displacements identified in previous chapters must not, however, be thought of in either-or terms: empty*, open* future potential has *dis*placed rather than *re*placed embedded, embodied, contextual and individualised futures that were pre-set by nature, fate and god(s). It means that earlier forms of being, relating and understanding have not been eradicated; they have merely been placed outside the modern public frame of reference, relegated to the private realm of contemporary existence where they have been rendered largely invisible. From this position in the shadows of public concern, however, they play an influential role in the formation of paradoxes that so powerfully mark the modern age. As such, these negated modes are implicated in the specific contradictions that arise when modern futures are engineered institutionally by political, legal, economic or scientific means and when they are transformed technologically. To understand these contemporary future-transforming processes therefore, requires that we perceive them together with their silenced 'other', whose countervailing forces undermine intentions, plans and promises: future making slides into future taking, progress into peril, intention into un-intended consequences. In the light of these tensions we begin the chapter with thoughts on Promethean power which is the underpinning theme for this and the next chapter of our exploration.

Promethean Power

Prometheus, a lesser Greek deity, stole fire from his fellow gods to give to the people. The gods, however, did not want to part with this procession because they thought humans not prudent and restrained enough to handle this precious gift with wisdom. Thus, when his deed was discovered, Prometheus was severely punished (Ferguson 2000: 69; Littleton ed. 2002: 151). When we use the term 'promethean' today we refer to an awesome power to set something in motion without an equivalent power to know and be mindful of potential consequences. In the course of these two chapters, therefore, we consider the complex and contradictory relations between progress and peril, future making and future taking, the frontier spirit and institutional irresponsibility*.

We explore the pursuit of speed to the point of stand-still in the present, and we examine the impossible quest for control in the modern context of human freedom and globally networked relations.

Much has already been said in previous chapters about the idea of progress and the fervour with which progress was pursued at a historical period when the political landscape was in turmoil and science became a dominant force in society. By the time the idea of progress had taken hold, people rather than their gods were in charge of the future. With this shift in ownership, the future became a social rather than a sacred domain. As such it became no different to space which, unlike the temporal sphere, had always been the domain of human action. Like any other territory which was subject to human design, planning, management and regulation, the future became a realm to be administered. This in turn brought forth new experts on the subject, in this case not experts who would predict what was going to happen to the lives and plans of individuals and groups but specialists in *producing futures to blueprints*, which meant achieving desired results in and for the present.

These experts in future making, who were primarily drawn from the ranks of science, politics, policy, law, engineering and economics, systematically applied the principles of their disciplinary knowledge to the technological and social engineering tasks at hand. Their underpinning understanding of how the world works, as we showed in the last chapter, was a largely mechanistic one: of objects that move in space, propelled by levers and pullies, and held to the ground by gravity. Theirs was a world of parts and wholes that operated along linear chains of causes and effects where each cause was thought to be proportional to its effect: a hard push moving an object further than a light one, a steep hill requiring more energy to propel a vehicle than a flat road. It was a universe of bits that could be assembled into functioning wholes, taken apart and re-assembled again without affecting the integrity of the object. Moreover, this system of parts and wholes in motion was amenable to counting, measurement and quantification. As such it became manageable, allowed for control and could be translated into money.

As we indicated in the previous chapter, this mechanistic way of understanding reaches deep into our cultural history. It dates back at least to Greek antiquity in the 5th century BC when Anaxagoras originated the theory of atoms in motion. As Adam notes in earlier work:

> Anaxagoras (fifth century BC) opposed Heraclitus' theory of change with a mechanical theory of nature and substituted the idea of opposing forces with

> one single cause of motion. Nothing is produced from nothing. Nothing is lost. Coming into being is nothing but different mixtures of the same, infinitely small, indestructible 'germs' which are the absolute, unchangeable essence of the universe. Equally, death or passing away is nothing but the separation of a particular combination of those elements of the eternal essence. There is therefore no change only movement, relocation, and recombination of the unit parts into different form. The material elements are inert, without cause or purpose. *Nous*, reason/mind/spirit, is the single moving and motive force that creates order out of chaos, separates the elements and sets the cosmos in motion. (Adam 2004: 25)

In its modern, mechanistic and materialist guise, this perspective still operates today as a powerful metaphor in our everyday understanding not just of physical but also of social and organisational processes and relations. Its principles continue to underpin many of our social efforts to control the future, create progress, and speed up development in a particular direction. Equally, the feasibility of strategies in politics and business, for example, is often decided on the basis of its underlying assumptions where interdependencies are acknowledged but thought to be amenable to designed intervention and transformation. When socio-technical processes* are understood as material objects in space then *details* come into sharp focus and the *function* within a system is fore-grounded, both of which advance knowledge for action. At the same time, however, we lose sight of connections, interdependencies and temporalities. In addition, this perspective concentrates attention on the present at the expense of history and futurity, with the effect that we are no longer able to grasp our products with reference to their social origin and socio-environmental destiny. Power becomes promethean in the sense that the power to act and transform is not matched by a capacity to know and be mindful of interconnections, implications and potential effects.

With hindsight it is possible to recognise that many of the most successful as well as the most problematic products of the pursuit of progress have their origin in just these interdependencies and relations: fossil fuel, plastics, nuclear energy and geno-technology being just some of the most prominent examples. Thus, the exploitation of fossil fuels and the invention of the heat engine brought untold social advantage whilst being accompanied by problems of pollution and resource depletion. The creation of plastics transformed our daily lives and afforded us with previously unimagined conveniences whilst also presenting us with new intergenerational health problems that today affect reproductive processes right across the animal kingdom, including those of humans.

The splitting of the atom and the production of the nuclear bomb offered not just new forms of energy production but also vastly increased powers of destruction. Nuclear power left us and successors with huge financial burdens for the essential management of waste products that remain dangerous for millennia. Its radioactivity affects all living things at the level of cells where it produces new forms of cancers that are bequeathed across generations. Genetically modified crops, finally, thus far fail to live up to their promises and refuse to behave to their laboratory design specifications: all interact with their environments. Some crops have their seeds carried by the wind beyond their allocated safety zone; others are assisted by bees and other insects which carry their pollen and begin the unplanned process of contamination.

All these technologies produce futures on a similar basis: all rest on or are built on mechanistic foundations. All come attached with enormous promises: they *will* solve the problem of resources and world hunger in a context of a vastly increasing world population. They *will* produce cornucopia and aid the potential for world peace. They *will* eventually provide cures for many of our most dreaded diseases. Importantly, in each case the knowledge to do, create and transform is greater than the capacity to know, mind and take care of long-term consequences. Finally, in all these future-making and future-transforming technologies the achievements cannot be separated from their shadow side of costs and negative effects. Let us now consider some of those promises in a bit more detail by focusing very briefly on nuclear power and genetically modified food.

'Electricity too cheap to meter' was an early slogan of the nuclear industry some fifty years ago when nuclear weapons technology mutated into production of peaceful energy. The promise was based on the faith that fast breeder reactors would be developed which could create their own fuel while generating electricity. Again, with hindsight it is possible to see that this promise could not be fulfilled: of the three fast breeder reactors ever built only one is still in operation and none managed to breed their own fuel (Hughes 2006: 46; Lidsky 1983; Montague 2005: 14–15). It also became clear that the industry had to be heavily subsidized out of the public purse, not just for its production but also its insurance and waste management. Finally, it became apparent that decommissioning costs were never adequately factored into the initial calculations. In the UK alone this estimated cost has risen within a few years from £56 billion to £70 billion and most recently to £90 billion (Morgan 2006). Due to concerns about cost, safety and security, nuclear energy ceased to be the favoured option in many countries: in the US the last new plant was built almost thirty years ago, in the UK it was almost twenty years ago and the last facility in Continental Europe was built ten years ago (the new Finnish plant underway at present being an exception).[3]

In response to wide-spread public distrust and political reticence, in 1996 the former president of the American Nuclear Society, Alan Waltar, launched the Eagle Alliance in order to "revitalise nuclear science and technology in America". In its membership brochure, the Eagle Alliance envisioned a world where "a safe, healthy, and sustainable society" is realised through the continued development of the "full potential of nuclear science and technology". The Eagle Alliance seeks to clarify for the public that "nuclear technologies, used in medical diagnostics and treatment, industrial processes, agriculture, food preservation, and energy, have proven beyond question to be a major benefit to all humanity". It projects a world in which science is fully dedicated to the service of humanity, reducing the distress of disadvantaged populations and assuring the blessings of a sustainable future for all peoples. "We believe" continues the brochure text, "that this vision cannot be realised without nuclear science and technology".[4]

Today, in light of concern about global warming, some of these promises return while others have mutated to suit the new context. Today, citizens the world over are assured once more by their political leaders that nuclear power can generate safe, secure, constant, unlimited supplies of electricity. Most importantly, and unlike power produced on the basis of fossil fuels, they are promised that the nuclear option provides 'clean' energy that does not contribute to the growing threat of climate change. It is therefore hailed as the answer to global warming. The problem with today's promises (as with earlier ones) is what is externalised in the calculations, on the one hand, and how the unknowable is handled on the other. To understand nuclear power as the answer to global warming necessitates that one ignores any calculations about the *contributions* to global warming which inevitably arise not just during the building of nuclear power stations but also during the very costly, toxic

[3] Here too however, the costs are already set to rise unpredictably. See Hughes' Nuclear Dossier in the "The Ecologist", June 2006.

[4] See Sullivan (1998: appendix to chapter VI). The brochure text is reproduced online at http://www.vanderbilt.edu/radsafe/9611/msg00220.html and http://www.eaglealliance.org/. See also Adam (1998: 194–209).

and polluting mining of uranium. Jon Hughes (2006: 43) cites a study undertaken by the Canadian nuclear industry which estimated that 1.6 million tonnes of steel and 14 million tonnes of concrete would need to be produced and then transported to the chosen site. Just to set this in context, one tonne of CO_2 is dissipated into the atmosphere for every tonne of cement produced. Even more worrying are the figures for the extraction and milling of uranium. Uranium is a finite resource which is estimated to run out within fifty years on current demand and in less than 20 years on the estimated increased demand.[5] Moreover, the mining of uranium is uneconomical in that it requires more energy for its production than it will generate and it is highly polluting, its vast quantities of toxic waste poisoning ground water and the atmosphere. As part of the production process, furthermore, uranium needs to be enriched which entails using half a tonne of fluoride for every tonne of uranium hexafluoride. However, the contribution of fluoride to global warming is nearly 10,000 times that of CO_2 (Hughes 2006: 47). This list of exclusions from the calculations on which the new promises are built by no means exhausts the numerous factors that have to be externalised, bracketed and considered irrelevant to the current debate before the conclusion can be reached that nuclear power is the best answer to climate change.[6]

A similar tale of selective accounting and incompatible positions between opponents and proponents applies to the debates about genetically modified food. Let us summarise here some of the high-profile public promises of that industry, which seem to be produced to a rather similar script to those made about nuclear power by Alan Waltar in the late 1990s (Adam 1998: 212–228; Adam 2000b). Genetic modification, it is argued by its proponents, increases productivity and thus has the potential to alleviate world hunger. It can help transcend agricultural limits set by weather and seasons and provide more nutritious foods. It can increase the diversity of foods available to us. Genetic engineering of crops and animals, it is claimed, improves on nature and increases bio-diversity. GM food can be stored longer; decay can be held at bay. Genetically modified crops can be resistant to diseases and pests while being tolerant to herbicides and pesticides. This, it is proposed, will

[5] The figures tend to vary in detail but the overall message is clear: uranium is a finite resource that is going to run out and the speed of is depletion is directly related to demand.

[6] For the continuing debate see for example Storm van Leeuwen (2005).

reduce the need for herbicides and pesticides (incidentally produced by the very companies that are developing this geno-technology with all its problem-solving and palliative powers). Despite these promises, it turned out that even over the short period since genetic modification of food first came onto the public agenda, the problems associated with it far outweigh the benefits: unviable organisms, disastrous crop failures, the decline of whole species negatively affected by the crops, such as the Monarch butterfly, and health problems in animal and human guinea pigs being just some of the more widely cited examples.

In both cases predictions failed to materialise. Promises were broken. Expectations were disappointed. Trust was abused. And yet, incredibly, the same palette of promises re-emerges for old and some of the new industries on the horizon, for example, in response to public unease about the emerging nano-technology industry. How can this be? Let us consider here some answers to this question on the basis of technologies we have already encountered above. The issues that emerge in response to this query have deep significance for understanding how the future is so consistently discounted* at the level of public and institutional concern.

In each of the cases the public is assumed to suffer from amnesia, incapable of remembering the last set of broken promises of the science-business-politics alliance. However, on the whole, people living in socially embedded chains of relations and dependencies *do* remember. The problem of social amnesia, short-term memory and lack of concern for long-term effects arises not in the socially embedded lives of people but emerges instead within institutional structures. This is the case because in their professional capacity none of the institutional actors operate in equally embedded contexts. Thus, for example, politicians act on behalf of citizens within a political framework and with a public mandate of four to five years. Business operates in the temporal context of an extended present where focus and orientation rarely endure beyond the next shareholder meeting and the schedule of quarterly results, and where new CEOs have their own exit strategy in place long before they begin to develop their restructuring plan for the company's future (Sabelis 2001). Dis-embedding and disconnection from chains of obligation that stretch into an extensive past and future, therefore, has to be appreciated as one key explanatory factor.

Attention to the underpinning functional requisites of techno-science allows us to trace some further associated yet different connections. Many of today's most successful technologies are founded, as we suggested earlier, on the disembodied, dis-embedded and de-contextualised principles of mechanistic science. The products of these sciences, however, do not exist in abstraction. Rather, they form an integral part of human social existence where they are appropriated, used interactively, and absorbed into daily life. The products have become naturalised as part and parcel of who we are, how we live, what we are able and unable to do (Latour 1993, 2004). Moreover, due to the complexity of modern life and the resulting division of labour, we cannot know the products of the pursuit of progress in all their facets. Instead, we interact with and use them on the basis of practical know-how and second-hand expertise alone (Weber 1989/1904–5). This means, in the daily context of lived techno-science, knowledge and know-how have drifted apart and the gap is still widening. This has problematic consequences and gives cause for concern. As Hannah Arendt argues,

> If it should turn out that knowledge (in the modern sense of know-how) and thought have parted company for good, then we would indeed become helpless slaves, not so much of our machines as of our know-how, thoughtless creatures at the mercy of every gadget which is technologically possible, no matter how murderous it is. (Arendt 1998/1958: 3)

Mechanical principles inappropriately applied to social systems and know-how without knowledge, therefore, constitute another explanation for the lack of appropriate responses to the politics of broken promises.

Two points arise in relation to the combination of dis-embedding processes raised above. First, the scientists, engineers and economists who are major players in the production of progress are not exempt from this division of labour in the sphere of knowledge where expertise is narrowing into ever-decreasing specialised niches. Here, professional and private knowledge spheres have come adrift and no longer map onto each other. In their private mode of being these specialists worry about how their creations might be used and to what perils they might lead. In their professional capacity they are pursuing knowledge for the sake of advancing knowledge or opportunities for the sake of enhancing opportunities. For example, use, application and socio-environmental consequences are not part of a scientist's professional remit. They are considered not a scientific but a socio-political problem. Whenever the scientists and the lay public are brought together to discuss new technologies and their safety—be this through such efforts as 'up-stream public involvement', 'see-through science', 'citizen juries', 'public involvement in science'—these separations of knowledge spheres and concerns rise to the surface (Kearns et al. 2006). It becomes starkly apparent that at the forefront of public knowledge and the institutional production of progress no-one is in charge. No-one takes overall responsibility. Nobody feels able or motivated to take the long-term view, no-one obliged to unite private and professional domains of concern.

Secondly, for today's public administrators and regulators difficulties arise when technological products are treated as material objects and abstracted from their sociality, that is, from their lived interdependency with people and nature where everything connects to everything else in seamless processes that extend into an open-ended future. As *social* things technological products are not mere artefacts in show-cases. Rather, they are socio-technical agents that produce new relationships and interactive effects, adding not just to the overall good but equally to the overall entropy of their system, its waste and its pollution. *The modern Promethean power is unleashed, therefore, where connections and interdependencies have been severed, where context and temporality are eliminated from the relevant frames of reference, and where moral concerns are considered out of bounds within the knowledge spheres that are at the heart of producing progress and its accompanying perils.*

Thus, in contexts where abstraction, disconnection, de-temporalisation and fragmentation of experience and knowledge abound, personal involvement, responsibility and long-term perspectives tend to be placed outside the frame of reference from which technologies are developed, tested and considered safe. Problems that accompany the successes of the era of progress, we need to appreciate, arise from those very displacements and repressions. When for example the safety of a nuclear or genetically modified product is established under laboratory conditions over a time-span of a few months or years, the same product placed in the environment is likely to produce symptoms over substantially longer periods in previously untested bodies and places, working their way through organisms and their environments into a boundless future. From within the institutional framework, the historical embeddedness of products with their interactively open future is bracketed. Consequently it is extremely difficult (but not impossible) to know ourselves responsibly connected to the eventual time-space distantiated* outcomes of our decisions, actions and inactions. Abstract, de-contextualised and discontinuous knowledge, short-term perspectives and the division of labour facilitate the production of promethean power almost by default, producing peril from progress. It is therefore worth our while to further investigate the relations between knowledge and know-how, progress and peril, so that we may explore potential openings for approaching them differently.

Future Makers, Future Takers

The taming of wilderness, opening up new frontiers for settlement and human activity, prospecting new resources, exploring the invisible worlds of the deep and reaching to the stars, all these are activities where the pursuit of progress is combined with a frontier spirit. The mind that moves and creates order out of wilderness and chaos today is the (economic-technical) rational mind of modern man and (significantly less so) woman. The roots and nature of this rationalism were elaborated in the previous chapter. Here we want to offer some illustrations of the kind of paradoxical effects that can arise when future-making is built on these foundations, show how future making almost imperceptibly slides into future taking and offer some thoughts on the relation between the two.

When we look at future making efforts such as the creation of safety, sustainability, salubrity and security, for example, we quickly note that they seem to have their opposites encoded at the very base of their method and approach. And we cannot fail to realize that, depending on the technologies involved, future taking comes in many different guises: futures may be spoilt, foreshortened or eliminated. Since the frontier spirit is pivotal to understanding those interdependencies we shall begin this exploration with short notes on this particular way of extending into the future.

The first thing to appreciate is that the frontier spirit is not an exclusively modern phenomenon but reaches back for thousands of years of human history. What sets ancient and modern forms of it apart is both the scale and pace involved, especially with regard to the unintended consequences of this particular form of future making. Today the scale of effects is global. Equally, the pace of consequences such as the depletion and degradation of resources has dramatically increased. Thus, resources that used to be exhausted over very long periods with the rate of depletion only just outstripping the rate of renewal—forests, fish stocks, grazing pastures and top soil, for example—are today disappearing in a few hundred years and even decades. Thus, if you take nutrients from the soil without returning equivalent amounts to it then erosion and degradation takes place. However it is the scale and pace that make a significant difference: one cubic meter of top soil with its ecosystem of bacteria and micro organisms, which took around 100,000 years to develop, is continuously depleted, eroded and/or salinated in less than a person's life time by industrial modes of agriculture. This means that today deserts grow where once there grew forests as part of delicately balanced eco-systems and interdependent plant and animal communities (Pimentel 1993; Pimentel et al. 1995). Importantly, of particular interest here are the contradictions associated with efforts to control and manage the future. Towards this end, examination of the frontier spirit can provide helpful insights into the paradoxes that arise with future making in both the ancient and the modern era.

Secondly, when new territories are colonised, we find that the dividing line between future making and future taking is difficult to draw. The conundrum arises whether or not those colonisations occur today, happened a few hundred years ago or date back several thousands of years, as in the case of Meganesian settlers (Flannery 1994).[7] The latter, for example, were deceived by their new territories' apparently unbounded resources. According to Tim Flannery, some of the first future takers emerged some 60,000 years ago in Australasia and Meganesia when populations grew too large for their tightly delimited supporting environments. He draws on archaeological records which suggest that some of these early future makers were settlers in new lands, thus unable to draw on collective past experience of the ecological interdependencies of the habitats they had left behind. "Without predators and surrounded by naïve prey", argues Flannery (1994: 160), "people would have become, in a sense, gods. For they were now all-powerful beings in the land of plenty". Lacking the benefit of accumulated collective wisdom, resources were used up faster than could be replenished and, despite great cultural developments, future taking became endemic among early settlers of the Pacific islands and the greater landmasses of Australasia, Tasmania and Australia.

A similar relation emerges from the story of beef, as told by Jeremy Rifkin (1994). Rifkin's insightful study of American beef culture takes us back some 6000 years to nomadic herdsmen that clashed with settled agriculturalists in what is today the Middle East, Europe and

[7] The story is told by Tim Flannery (1994), who coined the phrase 'future eating' for processes we identify as 'future taking' which accompany 'future making' and the transformative practices associated with social and environmental engineering.

the Indian subcontinent. At that time cattle were not only given as sacrifices to the gods but already appreciated as an essential wealth-creating commodity.[8] Beef eating and herding, of course, posed little or no socio-environmental problems as long as population density was low and grazing land in unlimited supply. Today those pre-conditions no longer apply. Instead, the world-wide production of some 1.3 billion grain-fed cattle, kept for meat consumption, is recognized to be one of the primary causes for desertification which is tied to four interlinked, environmentally damaging processes: deforestation, over-cultivation and compaction of soil, overgrazing and improper irrigation. Thus, for example, each animal consumes some 900 pounds of vegetation per month and compacts the soil with a pressure of twenty-four pound per square inch. The Worldwatch Institute have produced numerous calculations on the effects of this particular industry. Amongst others, that one pound of beef in the shops equates to 35 pounds of eroded topsoil (Rifkin 1992: 203). Moreover, it is not only the soil that suffers when cattle are raised for beef consumption. Both water and the atmosphere are polluted, degraded and thus denied as living and breathing spaces to future generations of beings.

The future taking associated with beef culture is furthermore unambiguously tied to the North American manifestation of the frontier spirit. It is connected to the Christian fervour of pilgrims migrating west on the one hand and the utilitarian quest to tame the wilderness and transform nature through human will and desire on the other. Salvation was the long-term goal while the frontier set the task and focused the vision on the immediate future. This meant, suggests Rifkin (1994: 256–7), that "Americans adopted a wholly new time orientation, becoming a kind of temporal nomad, living only for the morrow". In other words, the taming of that new world required and produced men and women that were "unfettered by tradition or sentiment, unresponsive to past alliances and obligations, cued to the utilitarian needs of the moment". As such, the frontier spirit chimed well with the modern pursuit of progress, economic growth, and market efficiency we discussed in the previous chapter. This fusion of perspective, approach and effort maps neatly onto the Enlightenment view of the world with its stress on

[8] According to Rifkin (1992: 2), cattle have always been a medium for exchange and "one of the oldest forms of mobile wealth". Moreover, ownership of cattle was tied to power and in India the Vedic word for war means "desire for cow" (Rifkin 1992: 36).

utility, rationality, science, mechanisation and economic efficiency. It produces a coherent perspective focused on subduing, colonising and *conquering in both space and time*. With the gaze firmly fixed forward into the promising future, the thrust of actions is one of pioneering adventure and, in some cases such as the colonisation of the American west, combined with religious fervour. For these colonizers action took priority over questioning reflection, daring over historically embedded social concern, the pursuit of efficiency and effectiveness over environmental considerations. As such, the frontier spirit dissolves boundaries, overcomes limits and vastly increases the temporal horizon of human activity and impact.

Clearly, the frontier spirit produces not only tremendous progress and advantages but also dire consequences: species are being wiped out. Aboriginal peoples are oppressed. Pollution, desertification and global warming are on a seemingly unstoppable roll. People the world over are losing jobs and livelihoods. Entire countries are thrown into spirals of unserviceable debt (George 1989, 1992). Looked at from a futures perspective on resources, therefore, we can see that the frontier spirit is not just producing futures but also consuming them at an unprecedented rate: futures are eliminated like in the Indian myth where the serpent eats its tail or in Greek mythology where Saturn devours his offspring.

Resource depletion and degradation, moreover, are not the only forms of future taking that accompany so many efforts of future making. The quests for safety, salubrity, security and/or sustainability often turn out to have opposite effects to those intended. This can be observed, for example, in the development and widespread use of monocultures, in forest clearance for crop production, in large scale water projects to secure irrigation, in the trials of pharmaceutical products, in bio-engineering and in military endeavours. In all these practices the negative effects often exceed the positive intentions: efforts to improve on nature through genetic modification, for example, may result in reduction of bio-diversity and diminished fitness in genetically modified organisms, their futures being imperilled rather than enhanced. Equally, when progress is pursued in techno-scientific, medical and economic spheres, salubrity may be endangered rather than improved. Similarly, security may be threatened rather than strengthened with military interventions as the recent wars in Afghanistan and Iraq amply demonstrate. Two technologies—plastics and nuclear power—will serve to further illustrate the underlying relations between future making and future taking.

In the short space of a century, the development of plastics and their subsequent widespread socio-economic distribution across the industrial and industrialising world has changed people's lives beyond recognition. Plastics are manufactured resources that have infiltrated every aspect of modern living. Their residues are today found everywhere: in water, soil and air, animals, plants and humans, in heavily populated as well as the most remote regions of our earth. As Theo Colborn's seminal research shows, at their inception plastics were hailed to be safe and the recognition dawned only slowly that they were accompanied by time-space distantiated effects. Their inertness, stability and durability, all characteristics that featured on the positive side of the balance sheet, turned out to also constitute their dangers (Colborn et al. 1996):[9] they facilitated their bio-accumulation across the entire food chain until, today, there is no place left untouched by their system-invading effects. Moreover, the damaging processes work below the surface, unseen, unfelt and undetected. As such they facilitate a death that "is slow, invisible and indirect" (Colborn et al. 1996: 147). This means the dangers associated with plastics lack the tangibility, immediacy and drama necessary to galvanise politicians and regulators into action. Like cancers associated with radiation, the unintended consequences of the world-wide permeation of plastic materials are trans-generational hand-me-down poisons. Unlike cancers, however, the damage is done not at the level of cells but the body's communication systems, which means that the immune, endocrine and nervous systems of animals and humans are affected. These poisons are passed on at critical stages of embryonic development but do not develop into symptoms until the embryos have matured into adults. As such, plastics do not kill but endanger salubrity across generations. Notably, repair is not possible because the body does not recognise its enemy. By the time the harm is recognised, the attendant massive curtailment of futures is irreversibly set in train.

When we compare the future making and taking associated with plastics with that of nuclear war technology, we find that in the case of nuclear weaponry futures are not merely spoilt or foreshortened but potentially eradicated. With the invention of the nuclear bomb and a stockpile of nuclear weapons that has the capacity to eliminate the human race many times over, in other words, continuity can no longer be guaranteed and the potential end has, worldwide, become an

[9] For a secondary analysis of the temporal relations involved, see Adam (1998).

ineradicable feature of our present. No longer merely individual, death has become a collective potential that requires not only individual but also social responses. Moreover, since the knowledge cannot be erased or undone, even disarmament cannot alter the fundamental contemporary condition of the potential end in the present. This potential end in the here and now applies not only to the entire human species but also to vast numbers of other life forms. *Finitude, therefore, is no longer an individual but a collective potential in the present and one of the most fundamental human assumptions—that the following generation(s) will carry on where we leave off—can no longer be taken for granted.*

In summary we can say that future making and future taking, its accompanying shadow, are not contemporary phenomena only. Today, however, the nature, quality, scale and pace involved have changed the activities to a point where they are barely comparable to earlier forms.

> Throughout human history, humans have risked the unknown, courting both success and catastrophe. What differs now is the stakes, the magnitude of possible mistakes. Our activities no longer involve just one village and its neighbour, one valley or the next. The scale of human activity means that these experiments engage the planet. (Colborn et al. 1996: 246)

The very characteristics of embeddedness and interdependence that we eliminate from our designs and bracket from understanding and debate, we suggested, are the features that produce contradictions, unintended consequences and unforeseen surprises. Transformation entails unintended consequences since all affected elements and threads of an interdependent reality cannot be controlled. The promethean aspect of power therefore cannot be avoided or eliminated. However, unintended consequences and unwanted effects can be reduced significantly if some of the central problematic approaches to the future are revised.

Futurity Redeemed

To achieve a reduction in the negative effects and paradoxical consequences of future transforming activities requires changes at a deep structural level of sedimented knowledge and historically established, taken-for-granted meaning. It necessitates that we reconnect what has come adrift with the modern pursuit of progress and that we dramatically expand our frames of reference in accordance with the potential effects of our actions. Three clusters of issues in particular have arisen in this chapter and will thus focus our attention here in the concluding part: embeddedness and interdependence, connectedness and social memory, processuality and futurity. In each case there is a need to redeem what been repressed and bracketed in the public domain of modern life to its appropriate place in the future making scheme of things. Understanding the paradoxes we suggested in the introduction to this chapter is an important step towards alternative ways of future making. Thus, when we permit ourselves to be discomfited by the seemingly inevitable contradictions, we are allowing ourselves to perceive openings for change in the interstices between intentions and the kind of unforeseen and unwanted impacts we identified above. Let us therefore re-visit the issues raised here and identify some potential openings for change by focusing on technological products and their effects, people and their future-making actions, as well as contrasting perspectives.

The social practices that drive modern progress, we argued in this and the previous chapter, tend to employ mechanistic principles. These produce an interpretation of the world as a collection of material objects in motion, abstracted from their context and disconnected from their embedded and embodied interdependence. Difficulties arise because *in use* the technological products of progress do not behave according to the mechanistic principles of their design. Instead, they stand in an interactive relation with people and their environments. They have time-space distantiated effects and many of these are of an unanticipated kind. In the process of abstraction and fragmentation, connections to these potential time-space distantiated effects of actions are severed, with the result that it becomes difficult to appreciate any personal implication in the outcomes of collective and/or public decisions. This in turn militates against taking responsibility. We argued that, in a context where processes and relations are no longer embedded in their temporal continuum, problems and perils tend to accompany the successes of modernity. Future making all too easily slips into future taking. The first corrective move therefore would seem to be an effort to re-embed and re-embody the products of progress in their temporal continuum and to understand them as social. This would mean that we re-contextualise our products of progress and our technological projects and know them not as dead matter but as interactive processes, that is, acknowledge them as quasi social things.[10] As such they can be appreciated not a-temporally but

[10] It means we appreciate that technological objects become absorbed into everyday lives where they constitute an integral and interactive part of the socio-cultural fabric

as temporally extended, thus with their futurity re-instated. The socio-temporal nature of things will be revisited in the next chapter where it is addressed in greater detail.

When we shifted focus from things to people we found promethean know-how that brackets contextual, temporally extended and embedded knowledge with the result that people are more capable of acting than of knowing the effects of their future making. We showed that in public, professional and institutional life social memory is explicitly and comprehensively excluded where projects have been severed from past interdependencies, collective experience, social memory and ancient chains of obligations. The professional operational realms of science, business and politics were particularly noted for their social amnesia. To counter this tendency would require that we first know ourselves and our actions in their historical context; secondly, that we extend our framework of concern across generations to encompass time-scales that are appropriate to the decisions and actions in question; and thirdly that we connect the contemporary present to the future presents of potential recipients of our legacies. Thus, for example, with nuclear technology or genetically modified organisms we would need to match our horizon of care and concern to the time scale of potential effects, thus making it appropriate to the materials in question. If thousands of potential generations of grandchildren are implicated in the ramifications of our future making then we need to ask first, whether or not these potential loved ones would want to be bequeathed this particular blessing with its attendant perils and secondly, what structures we would need to put in place to provide *them* with the wherewithal to deal with our products of progress should these turn into legacies of peril.[11] More attention than is given at present will need to be spent on devising structures that connect our present to theirs in a seamless web of trans-generational communication and care.[12]

of life. Plastics, electricity, cars and computers, for example, are all illustrations of quasi social things with futurity.

[11] This is listed as points four and five in the ethical principles laid down by the Nuclear Guardianship Project which state: 4) Future generations have the right to know about the nuclear legacy bequeathed to them and to protect themselves from it. 5) Future generations have the right to monitor and repair containers, and apply such technologies as may be developed to protect the biosphere more effectively. Deep burial of radioactive materials precludes these possibilities and risks uncontrollable contamination to life support systems. See www.nonukes.org/ngl.htm

[12] Current efforts in the USA to find means to send messages of danger across a period of at least ten thousand years are still being pursued. See the websites of the

One way to begin this process would be a shift in perspective from abstraction to interdependence and connectivity, which is the core of an ecological understanding of the world. From an ecological perspective everything connects to everything else in a seamless web of material, spatial and temporal interdependencies that encompasses not just human beings but communities of animals, plants and inorganic matter, that is, all that exists on this earth and beyond. Where the mechanical perspective decontextualises, fragments and separates, the ecological one seeks connections and relations within specific contexts. It is these latter features that have been so meticulously airbrushed from the frameworks of meaning that underpin the industrial way of life. Yet it is these negated connections and interdependencies which today re-surface in the unintended and unwanted consequences of the carefully planned and executed pursuit of progress. To lessen those unwanted outcomes and implications necessitates that we re-connect what has been separated in the course of scientific development: people, technology and the environment; mind and matter; socio-cultural wisdom and the public quest for progress; economic pursuit of profit and social chains of obligation, care and responsibility; the timescale of resource use and the timescale of depletion—as well as future oriented action, knowledge and ethics, to name just some of the more prominent disconnections we touched upon in this and earlier chapters. With the shift in perspective to interdependence it becomes possible to recognise our implication in the fortunes and disasters of unknown others the world over. It allows us to connect our choices of food, transport, leisure, insurance, banking and modes of saving, for example, to droughts in Africa, to the debt crisis in the majority world and the economic collapse of entire countries in the East and Eastern Europe, to the wars in the Middle East, to the trade in opiates across the world or to the floods in Bangladesh and other low-lying built-up areas.

Similarly, when we expand the temporal framework of analysis and concern, and connect what is so carefully preserved in separate compartments and administered in disconnected institutions, then future making and future taking are seen in a different light. We realize that progress is always achieved at the expense of someone else. From such an expanded

---

Rosetta Project (http://www.rosettaproject.org) and the Long Now Foundation (http://www.longnow.org).

temporal perspective we could then begin to ask appropriate *social* questions: knowledge for whom or what? Who is likely to benefit, who to lose out? Moreover, in contexts where, due to the immense complexity of the processes and the time-space distantiated effects involved, there is a lack of scientific evidence to provide appropriate predictions about potential outcomes, evidence-based knowledge can no longer serve as sole justification for action or inaction. Instead, socio-cultural wisdom, values and ethics need to form the base for decisions that affect present and future collectives. This in turn concentrates the mind on questions about responsibility: who and what are we responsible for? Who are we responsible to? And how far into the future do these responsibilities extend? Due to the temporal logic that underpins their respective professions, neither politicians, nor economists or scientists are in a good position to seek answers to these questions. This is an inescapably collective task that requires extensive public debate. Through the global commitment to sustainability, with its insistence that we cannot and must not separate the social from the economic and the environmental dimension of our actions, a start has been made to re-direct socio-cultural processes and projects in the direction of understanding connections and interdependencies. However, as long as their underpinning analysis is primarily rooted in spatial and material frames of meaning the good intentions cannot be brought to fruition.

Human affairs, as Hannah Arendt (1998/1958: 183) insists, exist in webs of relationships. In distinction to the fabrication of things, she suggests, action is not possible in isolation. Action is a temporally extended process and its products too are interactive, ongoing, boundless. Arendt (1998/1958: 190) suggests therefore that "the smallest act in the most limited circumstances bears the seed of the same boundlessness, because one deed, and sometimes one word, suffices to change every constellation". Thus, it is our capacity to act which produces processes that are irreducibly uncertain and unpredictable in their outcomes. We would want to argue in addition that this principle applies irrespective of whether we act (as individuals or collectives) or fabricate things that are used socially in the way we explained above. When we therefore seek to understand those social and socio-technical processes with mechanistic conceptual tools, we fundamentally miss the point. We make a category mistake that will inevitably lead to faulty analysis.

In our efforts to transform the future not just in a more predictable but also in a more responsible way, we therefore need to complement

Arendt's analysis with a temporal perspective that re-connects future oriented action, knowledge and ethics. Let us re-cap: fabrication produces material objects whose potential outcomes can be predicted in the present, based on knowledge of the past. This means all that lies between the creative action and its time-space distantiated effects remains not just *invisible* but is also considered *unreal* from within the knowledge framework of fabrication. If we want to connect products to their potential impacts, however, then the latency* of these products' effects needs to become an integral part of our understanding. To achieve this incorporation, we need to distinguish between the potential future *product* and the encoded *futurity* that permeates the present and foreshadows the future in the processes that are already in progress. *The decoding of this encoded futurity, however, cannot be achieved by prediction on the basis of a material past. It requires instead a shift in perspective from product as result to product as effecting process, that is, as ongoing becoming.* When we thus connect matter, space and time, then the futurity of organisms in general and of human beings and their projects in particular is redeemed. And, once futurity is reclaimed, we can find ways to connect responsibly to the time-space distantiated impacts of our actions and make the time scale of concern appropriate to the magnitude of our deeds together with their potential effects.

Reflections

Transformation is about control, about seeking to impose one's will on the world. Transformation of the future is about seeking to change what is assumed would have been there had no interference taken place. The technologies identified in this chapter are not just about taming or shaping the future, they are explicitly about transforming it. All are conceived as 'improvements' on nature. All fundamentally alter the present and future shape of things. The capacity to abstract and render static what is fundamentally ephemeral, moving and interacting has had tremendous advantages for *understanding* our world. For the quest to alter and *transform* an interactive, interdependent world, however, the abstract mode of knowing has turned out to be a tremendous disadvantage. It is the wrong conceptual tool for the job since, to insert creations conceived in the abstracting mode of knowledge into an ecological environment of give-and-take that extends from the beginning to the end of time is bound to result in unintended and unwanted consequences. To achieve such desired futures requires of transformers a thoroughly temporal and ecological mode of knowledge and operation. That mode however, alters not just how we transform the future but it also fundamentally shifts our understanding of what we consider important, relevant and justifiable.

CHAPTER SIX

FUTURES TRAVERSED

Introduction

As a result of treating the future as if it were space and/or matter a range of paradoxical effects arise. In this chapter we want to trace some of the complex and contradictory processes that are set in motion when the future is understood and approached as if it was a territory that can be colonised and traversed, or as a material resource to be used and consumed. Since the temporal world of processes is not like space and matter, we propose, the skills and approaches required to operate within and across it differ from those necessary for spatial exploration and conquest. As we traverse the temporal realm, for example, we do not just move within it but also tend to negate it, another meaning of traversal. That is to say, when we move across space, the territory that is being traversed remains; it continues to exist. When we traverse time and colonise the 'not yet', in contrast, the future is eliminated and transformed into an ever expanding present. It ceases to exist when the present becomes the exclusive operational focus. Moreover, our responsibility for the future becomes harder to keep in focus when transactions are conducted in and for this growing present, and thus gain meaning and significance solely with reference to it. In this chapter we focus on these interdependencies and show how important it is to take the future seriously, which means not just approaching it spatially and materially but also temporally. Focusing on the temporal and processual aspects of the future and integrating these with conventional spatial and material features, we increase complexity but, at the same time, also enhance the potential to avoid some of the unintended consequences that arise from the current capricious treatment of the future.

To better understand some of these interdependencies we first want to consider the paradoxical impacts of the compulsive pursuit of speed on contemporary efforts to transform and traverse the future. Here we show how the valorisation of speed* restricts the temporal perspective to the operational realm of the present. We follow this with an exploration of the modern quest for control. In previous chapters we have demonstrated how the future loses its reality status when it is abstracted from context and emptied of content: the pre-existing futures in which our predecessors believed evaporate and elude our grasp. Here we elaborate on cases where treasured plans and strategies for control disintegrate as actions conceived in the realms of space and matter get absorbed into the global process web of interactive give-and-take. In the last section of this chapter we consider the politics of posterity*, and scrutinise the temporal relations that currently characterise it. We demonstrate how a politics of space and matter founders in the temporal realm of future making.

Speeding into the Present

Speed provides evolutionary advantage. Applied to both animal and human life, for example, it is often vital to survival, be this for catching prey or escaping one's pursuer. Historically, for human societies it has enhanced military prowess and economic competition, and has thus been perceived as a generally good thing for individuals, groups and organisations. The contemporary valorisation of speed therefore connects our efforts with those of earliest ancestors, most recent predecessors and contemporaries the world over. Yet, when we explore the quest for speed from a futures perspective, we note some modern variations that significantly differentiate today's approaches from pre-modern and ancient ones. The foundations of our analysis of some of these differences have been laid in previous chapters. Here we want to build on these and develop them in a particular direction.

Progress*, we showed in previous chapters, is a product of human will, not natural necessity. It could be advanced, Hans Blumenberg (1986: 240) proposes, "by method, organization and institution, and condensed by speeding it up".[1] The speeding-up referred to here is intimately tied to an economic perspective on the future. In Futures Traded we explained how extraction from context and emptying of content made the future amenable to translation into money. Once time and the future are equated with money and traded as abstract exchange values, however, speed provides not only evolutionary and cultural but also commercial advantage. That is to say, when saving time also saves money, accelera-

[1] Quoted in Nowotny (1994/1989: 46).

tion becomes an economic imperative. Thus, once the time-speed-profit combination had been established and has begun to permeate every aspect of the industrial way of life, no social relation and no approach to that life could escape its influence: the social worlds of education, work and commerce being prime examples. Jeremy Rifkin (1987: 3–4) goes so far as to suggest that "the idea of saving and compressing time has been stamped into the psyche of Western civilization and now much of the world". To understand the depth of its socio-cultural penetration, we need to focus once more on the underlying rationale of the assumption that time equals money and speed equals profit.

As a resource money can be quantified, accumulated and saved. This is clearly not the case with the future. On the contrary, futures diminish, evaporate and get absorbed into the present as time is traversed and futures are colonised and used. When we get to a destination more speedily, therefore, that destination is no longer the future but becomes the present. For the metabolic systems of individual organisms, increased speed means approaching death faster. As we already indicated above, moreover, before the future could become an economic medium for exchange it had to be abstracted from context and emptied of content. That is to say, unique individual, collective and environmental futures had to be cleansed of their context-bound uniqueness before they could be traded as abstract exchange values. When we engage in speed practices, however, it is specific not generalised futures that are implicated and these cannot be saved or accumulated. In all its *social* (contextual, embedded and embodied) dimensions, therefore, the future—lived, generated and known—stands in a highly paradoxical relation to the issue of speed.

In Futures Traded we further identified the link between futurity, credit, investment and profit. From within that perspective efficiency means producing something or performing a task in the shortest possible time, since to spend as little money as possible on labour and other resources assures the fastest returns on investment. Here we want to relate the valorisation of speed to the speed-efficiency-profit equation and explore some implications for our relation to the future. To map these interdependencies we draw on work by the French political theorist and technology critic Paul Virilio,[2] who set out a number of historical distinctions between modern pursuits of speed on the basis

[2] Virilio (1991), (1995/1993), (2000/1999).

of three forms of technology: nineteenth century transport, twentieth century transmission and twenty-first century transplantation. While all three are means of speeding up the traversal of space and time, they also differ in the ways they create and negate futures. It is these similarities and differences that are of particular interest to us here. In particular we want to keep in mind the space-matter-time distinctions as these help to shed new light on the contemporary production of futures and its multiple socio-environmental implications. We will structure our discussion around Virilio's headings of 'transport', 'transmission' and 'transplantation', but our focus on the future will take the analysis in a rather different direction.

The underpinning assumption of the valorisation of speed is that saving time means saving money and thus increases efficiency. But does this relation hold? Do we save time? Do we save money? What does it mean to get to the future quicker?

Transport: Virilio argues that speed-based wealth is dependent on the speed at which people, objects and information can be moved across space. It is relevant to all sectors of society but has greatest significance for the military and commerce. With each advance in the speed of these diverse forms of mobility, the relation between time, space and matter is altered: increased speed either shortens the time involved or allows for greater territories to be traversed until, with today's air transport, people are potentially able to reach any place on earth within a period of two days.

Research on transport and speed has demonstrated a number of interesting interdependencies, all of which have long-term socio-environmental consequences. Two of these will serve to illustrate the point. First, it has been shown that the increase in speed of modern modes of transport has not resulted in massive time savings but that we travel more and cover greater distances instead (Brög 1996; Whitelegg 1993, 1997). When approached from a futures perspective, this finding relates to some interesting and surprising interconnections: since we do not tend to save time we also are unable to utilise the promised and/or expected saved time for future activities and translation into money. Similarly, we do not wrest from that traversal of space a future that could be invested or traded for other futures. Moreover, a future that is reached faster is no longer the future but the present. It is therefore a future lost. Furthermore, speed requires energy (Adam 2001; Hillman and Plowden 1996). Higher transport speeds therefore necessitate increased energy consumption. This depletes natural, mostly non-renewable resources, increases levels of pollution and advances global warming. All these are unplanned effects of the pursuit of speed that extend into the long term future: the more a competitive advantage in exploiting the future is sought by speeding up, the less it seems to be attained. To counter these unintended consequences, long-term future losses are offset against the hope of present gains. The future, however, is closing in and as such it can no longer be ignored in the present.

Unrelated but equally important is the effect of high speed on the capacity for extended vision and by implication the long-term perspective. When we walk, for example, we are able to look around, take in the landscape or the urban vista. Our vision is unbounded, our perspective open. With increasing speed our vision and perspective are progressively reigned in, forced into an ever narrower space, until attention is exclusively concentrated on the moment, the immediate space and the task at hand: the motorway, the cars in front, behind and to the side of us. In other words, the higher the speed the more the immediate space and time become all-absorbing. This is a troubling relation: while high-speed travel produces ever greater time-space distantiated environmental effects our perspective is progressively narrowed to the here and now. Stephen Bertman (1998: 2–3) calls it the "power of the now" which "replaces the long-term with the short-term, duration with immediacy, permanence with transience, memory with sensation, insight with impulse".

Coming back to the questions we asked earlier, we can see that the relation between speed, money and efficiency is a complex one when viewed from a futures perspective: first, the higher the speed of travel, the greater is the negation of time and futures. This means *de*creasing opportunities for trade of futures and translation into money. Secondly, while the valorisation of speed casts ever longer socio-environmental shadows, our temporal perspective progressively shrinks to only encompass the present. Environmental theorists work with the image of our 'ecological footprint' to explain spatio-material relations. This, however, refers to a space- and matter-based relation. It is silent with respect to extension in time and potential reach into the future. We propose therefore the idea of a *timeprint** to encompass the futurity of socio-technological products. The idea of a socio-environmental timeprint not only provides a temporal equivalent to the footprint but also illuminates in a graphic way the mismatch between timescales of impact on the one hand and socio-political concern on the other. Moreover, it enables us to recognise that the socio-environmental timeprint tends to grow in inverse proportion to our capacity to encompass and take care of the domain we thus occupy. Thirdly, since futures cannot be saved or accumulated the speed-profit-efficiency relation does not hold either. Such effects and relations are even further accentuated in the case of modern transmission through information and communication technologies (ICTs) where space has been rendered irrelevant and communications are conducted in the ever-shifting now of 'real time'.

Transmission: Over forty years ago Marshall McLuhan argued that the electronic world of ICT is a domain that fundamentally reorders the relation between matter, space and time.

> During the mechanical age we had extended our bodies in space. Today after more than a century of electronic technology, we have extended our central nervous system itself in a global embrace. (McLuhan 1973/1964: 12)

When instantaneity* and simultaneity* are achieved in information transfer across distance, space and the movement of bodies are rendered irrelevant to the communication of information. They become obsolete. This in turn has consequences for the senses and for the forms of communication that predominate in electronic information transfer. Thus, for Jeremy Rifkin,

> Electronic technology represents the final disembodiment of the senses. The more intimate senses, smell and touch, are eliminated altogether. Sight and sound are disembodied by machines, turned into invisible waves and pulses, transported over great distances with lightening speed, and then reembodied by other machines in the form of facsimiles, artificially reconstructed versions of the originals. (Rifkin 1991: 238)

With this technology an unbridgeable gap is opened up between the speeds at which information and physical bodies respectively can move across space: a discrepancy that can be as great as that between the speed of light and the pace of walking. Today, such gaps are routinely incorporated into the anticipations, plans and actions of members of industrial and industrializing societies, whether these involve travel, satellite television, e-mail and searching the world-wide web or relate to the movement of troops and equipment to scenes of modern warfare that are subject to electronic scrutiny and surveillance.

From a futures perspective we note that ICTs have extended the present spatially to encircle the globe. This global present in turn has negated the informational future that previously existed between the sending and receiving of a message. To write and send a letter creates a future in the present. In doing this, we anticipate the recipient reading it and writing a response. This simple activity therefore involves us in a constant movement back and forth between present futures* and future presents*. It is this movement, which forms the basis of many everyday activities, that is made impossible when through technology all focus, effort, power and prestige is invested in the now. Much of the temporal structure of human interaction at a distance, and hence the meaning of this interaction, is fundamentally altered when duration is eliminated and instantaneity operates across space.

> The power of the now is the intense energy of an unconditional present, a present uncompromised by any other dimension of time. Under its all-consuming power, the priorities we live by undergo transformation in a final act of adaptation to electronic speed. (Bertman 1998: 2)

Thus, for example, instantaneity is traditionally the operational realm of face-to-face interaction. For communications across distance, in contrast, waiting times had to be calculated into the communicative process. With ICTs this is no longer the case: instantaneous communication can be achieved irrespective of the distances involved. With electronic communication the duration required to traverse space is compressed to near zero. When established modes of communication are so fundamentally altered then human culture operates, as Zygmunt Bauman (2000: 128) points out, "in unmapped and unexplored territory, where most of the learned habits of coping with the business of life have lost their utility and sense." With the elimination of waiting periods, for example, the time to think, reflect, reconsider, plan and strategize has been substantially curtailed.

The public effects of instantaneous communication across great distances are graphically demonstrated in Stephen Kern's (1983: 259–285) historical work on the politics of war and peace during the early part of the twentieth century. Kern shows the dramatic impact of instantaneity on the diplomacy associated with World War I. With the telegraph and telephone, established modes of conduct had been forcibly altered: distant events had to be dealt with rapidly, requiring immediate responses with little or no time left for reflection, extensive consultation, consideration of other options, or for tempers to cool down. The new electronic context of instantaneous communication, where learnt behaviours and routines were no longer appropriate, required actions for which there was no precedent and responses for which there was no established code, leading to World War I and the hitherto unimaginable loss of life. Moreover, as Kern explains,

> There was not just one new faster speed for everyone to adjust to, but a series of new and variable paces that supercharged the masses, confused the diplomats, and unnerved the generals. (Kern 1983: 268)

Today, these multiple paces have become routine and the present has become the (almost) exclusive realm of communicative operations, with the future of formerly anticipated replies transformed into an outmoded relic of the past. Only when the technology breaks down are we reminded of this past and then the enforced confrontation with the future becomes a source of great irritation. Thus, when distance is rendered irrelevant for communication and instantaneity becomes the norm, the future as unfulfilled realm of expectations has increasingly negative connotations.

In contexts where communication is not only instantaneous but also *networked* across space the issues raised by the compression of duration are further amplified. The stock market is a case in point where information transfer is conducted not just *instantaneously* but also *simultaneously*. The significant consequences of this combination of temporal compression and spatial expansion for planning and predicting the future were demonstrated by the dramatic collapses over the last decade of banks, financial services and, indeed, entire markets in the East. When everything is instantaneously interconnected and simultaneously interdependent then assumptions based on material objects in motion no longer apply. Processes become unbounded. The unique moment disappears. Communicated information loses its location: it is both nowhere and everywhere. Ephemerality and transience re-emerge with a vengeance, negating hard-won certainties and stabilities which had been wrested from the uncertain future during earlier historical periods with the aid of rituals, social rules and institutions we described in Futures Tamed. With networked ICTs that operate in a temporal context of both instantaneity and simultaneity, traditional relations and approaches to the future are unsettled. That is, well established values of continuity, preservation and conservation become problematic at best, obsolete at worst. We see here an intensification of the relations and interdependencies we outlined with respect to the valorisation of speed in transport.

Once more, knowledge practices* rooted exclusively in the realms of space and matter lose their relevance and grip. Laws and regulations based on the spatial order of territory and sovereignty become unworkable. Power becomes mobile, shifting and slippery. Yet, in the domain of business, practices are already adapting to this particular contemporary condition. Thus, Zymunt Bauman (2000: 12) identifies a modern business elite that is in tune with these changed social structures where not durability but transience, not solidity but lightness, not space but temporality are valued. In "Liquid Modernity" he analyses this new realm of networked instantaneity against the backcloth of the solid world of material production and gives extensive thought to its characteristics and underlying principles.

> Indifference to duration transforms immortality from an idea into an experience and makes of it an object of immediate consumption: it is the way you live-through-the-moment that makes that moment into an 'immortal experience'. The boundlessness of possible sensation slips into the place vacated in dreams by infinite duration. Instantaneity (nullifying the resistance of space and liquefying the materiality of objects) makes every moment seem infinitely capricious; and infinite capacity means that there are no limits to what could be squeezed out of any moment—however brief and 'fleeting'. (Bauman 2000: 124–5)

In such a context concern with the long-term is rendered hollow. Moreover, when dependence on things that endure becomes a liability and a sign of deprivation, the quest for solidity and permanence becomes an anachronism. Transformative thought and action, as we showed in earlier chapters, was based on a deep confidence in a future that was amenable to human will and design. With the global establishment of instantaneity and simultaneity this confidence in the planned future becomes misplaced and displaced by the ever-changing present as primary focus of gratification and concern.

Transplantation: when we consider a contemporary technology of transplantation, such as genetic engineering, we find once more the relations between time, space and matter altered and approaches to the future transformed in the process. As is the case with ICTs, genetic modification reduces waiting times and displaces the future by massively expanding the present. In distinction to transport and ICTs, however, *genetic modification traverses not space but matter (bodies and organisms) and time.*

The movement of genetic material from one species to another, for example, is only possible because *all* living organisms share over 90% of their genetic material. It is this shared genetic base, which extends back for hundreds of millions of years to the beginning of life, that allows, for example, genetic material from an arctic fish to be spliced into that of a tomato. By decoding, comparing, excising, splicing, recombining, transferring, and cloning individual genes and sections of DNA, scientists are able to inject characteristics that are not part of the hereditary genetic make-up and to combine morphological and functional characteristics that have evolved separately for millions of years. With the development of these radically new sets of techniques, therefore, scientists are traversing time back to the origins of life. The ensuing negation of the millennia of separate and singular evolution that are being traversed allows for scientific intervention within the very basis of organic life. Moreover, this traversal of time extends not just to the beginning but also to the end of time, although the latter does not form part of the deliberate design. That is to say, once genetically modified organisms are reinserted in the environment, the ensuing interactive and reproductive processes facilitate an open-ended process that potentially extends to the end of time. With twenty-first century transplantation, therefore, our timeprint encompasses *all of time*. However, neither our knowledge practices nor our institutional structures are adequate to the responsibility that accompanies such temporal extension.

For our focus on speed and the future, it is the dramatic time saving achieved by contemporary genetic engineering which is of significance here. What differentiates these new genetic techniques from established traditions of selective breeding is the capacity to effect change in the present where conventional breeders had to await results over many generations of reproductive subjects. The new techniques of time traversal mean that millions of years of co-evolution can now be circumvented and reproduction cycles dramatically speeded up or cut out altogether. In a social system where time is money and the future is traded for its economic exchange value, this unprecedented acceleration of processes holds out the promise of enormous profits. Since, however, the system processes* of ongoing interaction are unbounded, open-ended, long-term, time-distantiated,* and often marked by extended periods of immanence and latency*, the scale of potential financial gains is matched only by the associated potential for unintended and unforeseen consequences. In the case of transplantation by means of genetic engineering we can conclude, therefore, that the creation of future uncertainty and indeterminacy is correlated with the power to traverse and compress, thus negate, time.

Looking back over the issues discussed so far we can say that in all three forms of technology—transport, transmission and transplantation—the speed-efficiency-profit relation changes with our focus on futurity. That is to say, conventional understanding in terms of space and matter is altered significantly when time and the future are moved into the foreground of our attention. From a futures perspective we see that getting to the future faster does not provide us with more but *less* future. The future is lost and replaced by an ever expanding present. We appreciate further that extension of the present to the furthest reaches of an emptied future means that the future loses much of its open character: the passage to it becomes ever narrower when much of the future is already used, disposed of, borrowed or spoken for. Options are being dramatically reduced and the potential for actions significantly curtailed. We recognize that increased control in the present associated with the traversal of space and time has produced a matching loss of control over outcomes, an issue to which we shall return in the next section of this chapter. Equally, we begin to realize that increase in speed seems to be accompanied by a proportional decrease in the scope of our visions. While our actions reach ever further into the future, our perspective and concern continue to contract to the operational realm of the present. This means in turn that when increasing speed of change displaces the open future with an extended present then the processes of innovation and waste production which used to be marked by linear succession all "crowd into the extended present" (Nowotny 1994/1989: 71). Speeding into the present, we can therefore conclude with Helga Nowotny (1994/1989: 49–50), "is filled with conditional negatives" and marked by limits and finitude. In metabolic terms, for example, increased speed quickens ageing and shortens the period before death. In socio-environmental terms speed requires more resources and produces more obsolescence. Instead of offering unending potential, therefore, contemporary knowledge practices that propel us into the present confront us with the possibility of individual, collective and environmental dead ends. Finitude at all these levels has become palpable.

This raises questions about efficiency and sustainability. Efficiency, as we have shown, is achieved by speeding up processes, that is, by compressing duration. A point is eventually reached with ICTs and genetic engineering where operations are conducted in the real time* of extended presents. Here, the gaze becomes firmly fixed on the immediate horizon and short-term material gains. It entails accelerated production and consumption of nature's resources in a context where traditional bonds of obligation, commitment and responsibility between generations have been severed.

> Efficiency is a present-oriented temporal value. Its concerns are purely instrumental. What counts is increasing output now. The past and future are seen as impediments to the full use and exploitation of the present. (Rifkin 1991: 267)

Since the valorisation of speed is a central feature of the industrial way of life and its operational logic, regard for the long-term future or concern for the wellbeing of future generations of humans and fellow beings becomes a contradiction in terms. Speed-based efficiency and sustainability, we can therefore conclude, are incompatible. However, it does not follow that sustainability requires a 'return' to pre-modern social relations. Rather, it depends on a willingness to remember social arrangements that were conducted in a temporal realm that extended from origin to destiny and operated on the principles of indebtedness, obligation and responsibility, as outlined in Futures Tamed. Clearly, there is and can be no going back, but by recalling 'memories of the future', that is, by remembering the visions and achievements of predecessors, we are able to recognize the importance of embodied embeddedness that connects us not just with ancestors and future generations of successors but locates our actions and inactions in a seamless web of environmental interdependence that reaches all the way to the birth of stars and an indefinite future. The implications of such an exercise in remembering past futures are significant for the issue of control, to which we turn next.

Blindfolded at the Controls

Control of the future had first been achieved through mechanistic science. This success at shaping the future to human design, as we showed in Futures Traded, was founded on a cluster of principles which included abstraction and de-contextualisation, quantification and spatialisation as well as the exclusion of both temporality and futurity from its methodology. Moreover, it entailed focusing on products rather than processes*, that is, the *outcomes* of change and action rather than change and action itself.[3] Control was further enhanced by *bounding* objects in time and space which involved denying their futurity and temporal depth. This meant placing process and connectivity outside the scientific frame of reference, thus treating materials *as if* they were bounded, static, a-temporal, and insisting on the irrelevance of context. The pretence, however, can only be carried so far given that our technologies are imbued with

[3] For an extended discussion on this subject see Prigogine and Stengers' (1984) work on the difference between mechanical dynamics and thermodynamics.

social values and knowledge practices, are caught up in a system of irreversible energy exchanges and leave legacies for an open future, as we have shown above and in previous chapters.

In Futures Transformed we first suggested the need to understand technology in non-technical terms, that is, not as bounded objects in space, fabricated to a blueprint, but as quasi-social things that stand in an interactive and transformative relation to their creators and users as well as to their environments. When we want to understand how control has been achieved and lost in modernity the distinction between a purely technical and a social understanding of technology becomes crucial. This understanding in turn relates to the way the future is approached, that is, whether it is treated primarily in spatial and material or also in temporal terms.

Let us take an example from nuclear technology to illustrate the point. With the splitting of the atom the bounded energy of the stars has been released from the invisible depth of existence. The beginning of time and the end of time are bridged in this moment of concentrated power. The enormity of the power unleashed in the moment of control, however, evades control, because the interactive process is set in train for an indefinite, open-ended period into the future. Furthermore, the future thus created is also the future negated: it is traversed in the full meaning of the word once the potential end in the present has become an inescapable condition of modernity, as we described in Futures Transformed. A number of interdependent socio-technical parameters contribute to this condition: first, knowledge of how to create a nuclear device is irreversible. It constitutes an integral and non-eradicable part of our world. Secondly, the world-wide stock-piling of nuclear material continues despite global treaties and efforts to reduce the potential for overkill. Thirdly, the number of countries with nuclear capability is still growing regardless of whether or not they are welcomed by the established members of the nuclear club. This creates a context where continuity can no longer be taken for granted. *The potential end in the present has therefore become a fundamental condition of modernity whilst control over outcomes has become synonymous with wishful thinking.* When the conventional knowledge system with its assumption of control clashes with the actual workings of its products, therefore, paradoxes arise and unintended consequences accumulate unchecked until society wakes up to the realization that neither its scientific predictions nor its methods of control are appropriate to the contemporary condition. Theo Colborn and her collaborators (1996) use the imagery that we are flying blind, hurtling towards a future with no-one in control, while Zygmunt Bauman (2000: 56) simply states that we are passengers in a jet plane that has no pilot.

Our 'blindness' derives from a reliance on knowledge practices that are out of sync with the social nature of technological products and thus make controlling them and managing their consequences difficult. We use the conceptual tools of abstraction, decontextualisation, spatialisation, quantification and bounding for the control of interactive, future-creating and/or networked technologies, such as hormone disrupting chemicals, radio-active materials, ICTs, genetically modified organisms or emerging nano-technologies. Thus we draw on mechanistic and deterministic materialism for projects and processes marked by interiority, latency, invisibility and time-space distantiated effects. This approach is criticised by Michel Serres (1999/1982: 108) as an unwarranted "metaphysics of the solid". Moreover, we are attempting to predict, control and regulate innovation, which is produced at an ever-increasing pace, on the basis of past-based knowledge. Here Aurelio Peccei (1982: 10) points out that reliance on knowledge of the past becomes inappropriate when "the future will no longer be a mere *continuation* of the present but a direct *consequence* of it". We are relying on knowledge of the past to control situations where the extremely long-term processes and time scales involved create fundamental indeterminacy. A prime example here would be radio-active waste management that draws on past-based geological data and risk assessment to establish for thousands of years into the future the safety of geological sites for the burial of radio-active materials (Shrader-Frechette 1993). Furthermore, we apply methods of control established under carefully-managed laboratory condition to open-ended interactive processes—genetic engineering would be a prime example here. In addition, we seek to control technological processes marked by instantaneity and simultaneity with tools designed for the behaviour of matter in space and its causal and sequential processes, producing what Alfred North Whitehead (1929) called fallacies of misplaced concreteness and of location. Finally, networked operations in 'real time' escape our grasp because control requires a gap between action and outcome on the one hand and linear succession on the other. With ICTs neither the gap nor the sequence are available for the insertion of controlling action. These are contexts where solidity evaporates, where "all that is solid melts into air", to use Karl Marx and Friedrich Engel's (1967/1848: 224) evocative phrase.

Time, space and matter are fundamental dimensions of social life. When we take one dimension out of the equation, as we have shown in this and the previous chapter, we run into difficulties. Contradictions and unintended consequences blossom. Management and control become an unrealisable dream. Here we identified some of the pertinent mismatches between knowledge practices and technologically constituted process-interdependencies. Once these are brought to the forefront of our attention, it is no longer surprising that the successes of mechanistic science are intimately bound to its excesses and the problems that confront contemporary societies the world over. We begin to understand why paradoxes abound and why, everywhere we care to look, increases in mastery are accompanied by loss of control.

These multiple temporally constituted tensions seems to generate a generalised sense of disquiet about responsibility for our actions: how to dispose of nuclear waste safely and responsibly thousands of years into an unknown and unknowable future; how to change the direction of energy policies to avert a worsening of climate change over geological time scales; how to secure food supplies for rising populations without worsening conditions for unlimited generations of people and fellow beings into an open-ended future? When decisions taken in the domain of politics and policy have implications that stretch over such vast time scales it becomes appropriate to talk of a 'politics of posterity'.[4] The tensions and contradictions associated with this domain of knowledge practice occupy us in the third and final section of this chapter.

Politics of Posterity

In the political sphere, the issues we encountered above re-appear. Here we find that the system of liberal democratic politics has developed historically as, primarily, a politics of space and matter. Its sphere of responsibility extends to a nation's territory, its resources and its wealth distribution. It is in charge of things that can be measured and counted: territories, people, institutions, traffic, crime, budgets and Gross National Products. With political debates on climate change, the management of nuclear power and its waste products, the regulation of chemicals,

[4] For an illuminating overview of how the extension of human political concern is necessarily connected to advances in technology, see Anderson (1987).

strategies about genetic engineering and approaches to nano-technology, however, politics has entered the future worlds of tens, hundreds and even thousands of generations hence. This means that decisions made and policies established by today's liberal democracies operate outside the spatial and material framework for which they had largely been established.

For the production of long-term futures liberal democracies draw on three dominant institutions: science, economics and law. As we argue in previous and subsequent chapters, all three knowledge practices are rooted in time-space logics that make their suitability for the task of guiding future-creating policies questionable. As we have seen, mechanistic science takes its evidence from accumulated knowledge of past and present matter and space. It consequently treats the future as both immaterial and unreal. Economics operates largely from the present for the present. Its forays into the future, therefore, tend to be parasitical on successor generations of humans and fellow beings. It treats the future as a resource like any other, consequently making a category mistake whose effects ripple through the entire system of instrumental economic action. Law, finally, is guided by precedent and arbitrates future operations on the basis of past and present matter, space and social relations. None of these dominant knowledge systems of contemporary liberal democracies, therefore, are fully equipped to deal with the futures of their making, and are thus limited in their contributions to the understanding, administration and regulation of the temporal realm.

In addition to this first set of difficulties associated with the impoverished futures competencies of the dominant institutions that guide contemporary politics of posterity, the reach of the actions undertaken by liberal democracies far exceed the period for which representative governments are elected. The latter's bounded terms of office and the fact that some voters are not yet born make the politics of posterity hugely problematic. Potentially the impact of *all* political action extends beyond a government's period of office. That is unavoidable. However, for actions that affect us and our children in the near future, there is an implicit understanding that the public have given a mandate to the government of the day to act not just on their behalf but also on the behalf of their children. With today's political decisions that affect the very long-term future this is no longer the case, since effects of policies are not just experienced by voters and their children but by an open-ended chain of generations without vote, voice or advocates to speak for them, nuclear power being a prime case in point.[5] Without institutional structures that encompass the operational realm of the future and without knowledge practices that can accord reality status to 'futures in the making*', today's future-creating politics tend to be conducted in both a *political* and a *knowledge vacuum*. This has serious consequences.

When risks and hazards, created within the jurisdictional time-space of a particular liberal democracy, transcend the boundaries of its legitimate authority, their impacts and costs are in effect externalised to other nations and/or to successor generations. The problems are shunted along, moved outside the sphere of responsibility. From a spatial and materialist perspective, hazards externalised across time are no longer recognised in principle as the concern of the offending nation's representative government in office. The long-term policies routinely pursued by contemporary liberal democracies, therefore, transgress the temporal boundaries of their political mandates and realms of jurisdiction. Moreover, since elected representatives are responsible to their electorate only, and since it is this electorate that bestows legitimacy on a government, the rights of people distant in time who cannot enact that power relation are 'discounted' in a way that is analogous to the discounting of the future* in economic processes. To put it differently, *the politics of space and matter operate with impunity in the temporal domain of the future in which all of us are trespassers.*

Without mandate for their temporal extension into the future, policies of liberal democracies are enacted in the frontier spirit* that we described in Futures Transformed. On the one hand, they are cut off from socio-political chains of obligation, chains that stretch back into the historical fog without cut-off date and definable beginning, as we explained in the preceding chapter. On the other hand, they are disconnected from a sense of obligation and responsibility extending into the future as far as the effects of decisions, actions and inactions are going to reach. It means that political representatives that act on our behalf face the same problems that befell settlers in a new land. In Futures Transformed we connected these difficulties to newcomers acting as free agents in a land of apparent unlimited potential. Carried by the frontier spirit they moved from one amnesia-afflicted action to the next

[5] See Shrader-Frechette (1993), and Adam (1998: ch. 6).

without the benefit of historically embedded memory. Further, their desires were un-tempered by the collective wisdom that bounds actions in the every-day realm of responsible community-based social relations that extend temporally into open pasts and futures.

The inappropriate assumptions and associated socio-political practices amount to *institutionally constituted irresponsibility**.[6] To begin to envisage and institute a politics appropriate to our situation therefore requires changes at the level of individual, collective and institutional action, knowledge and ethics. Some of these will be addressed in the chapters that follow. Here we merely want to stress that the politics of space and matter needs to be expanded to encompass the temporal reach of today's decisions, actions and inactions. This entails the creation of political structures suited to the timeprint produced by their policies. It means finding ways to encompass futurity and processes that extend beyond the present and to embrace the lived and living futures* that constitute an inescapable feature of our lives and are implicated in everything we do, as we show in the next chapter. Finally, as citizens we need to acknowledge that our collusion with the policies produced by political representatives makes us responsible for the techno-futures set in motion: yesterday, today and tomorrow. We are charged therefore as citizens, professionals and private individuals not just to understand the contemporary bracketing of futurity but also to seek openings for change that help reconnect the spheres of social action which have come adrift during the scientific age: knowledge, action and ethics. Where knowledge about potential future effects is unobtainable, plans and actions needs to be judged and adjudicated on the basis of ethics. Ethics too, however, is in need of re-orientation, as we show in following chapters, if it is to become appropriate to the contemporary condition.

Reflections

The temporal realm of futurity is not to be confused or conflated with the domains of space and matter. It has different 'properties' and our interactions in that realm produce results that differ significantly from ones that mainly affect space and transform matter. Where the future is approached as if it were space or matter, surprises ensue and paradoxes

---

[6] See Beck (1992/1986) and (1999) where he too comes to the conclusion of structural irresponsibility, if by a different analytical route.

arise that are difficult to handle with the conventional knowledge practices that dominate the institutional spheres of contemporary industrial societies. Like space, the future can be traversed and like matter it can be used and consumed. The effects of such actions, however, are fundamentally different where the future is concerned. As time is traversed the future is transformed into the present, thus ceases to exist. By contrast, when the future is used and consumed in the present, it nevertheless continues to extend temporally, producing effects for future generations. By creating long-term futures we appropriate the present of successors. In this case our timeprint exceeds our allocated operational domain and established sphere of responsibility. This means we operate as uninvited migrants and trespassers in the temporal territory of successors.

Efforts to transform, traverse and control futures tend to result in unintended consequences, and paradoxes seem to bloom proportional to the reach of the actions involved. Thus, for example, speed progressively forces attention on the present, making it ever more difficult to extend our vision to encompass the temporal depth and breadth of our making. With nuclear technology our impact extends to millennia whilst confronting us with the potential end in the present. Economic instrumentality, finally, treats the future as a free resource, ignoring that it therefore exploits the future presents of others not yet born who cannot hold us to account or charge us for its use and/or depletion.

Using inappropriate conceptual and ethical tools is one way to ensure that paradoxes and surprises continue to accompany plans, strategies and actions. The task, therefore, is to make those tools more appropriate to the contemporary situation. Accordingly, the last part of this book will attend to those tools and scrutinise them as bases for such renewal. This is not to say that such revision will solve all uncertainty. It will not. But it will help us to distinguish genuine indeterminacy from that arising out of the use of inappropriate tools. Where non-knowledge predominates, we are in the realm of morals and ethics. Here decisions have to be reached and quandaries have to be adjudicated by collective deliberation on the basis not of evidence and prediction established on past facts but on *what is right and just*. Knowledge, action and ethics, finally, will need to be aligned flexibly. They need to be combined according to the requirements of specific contexts of the future-producing practices in question.

CHAPTER SEVEN

FUTURES THOUGHT

Introduction

Until now, we have focused on two interlinked aspects of how humans construct and relate to futures: *knowledge practices**, and the *implicit assumptions about the future* that underlie them, linking diverse practices to one another. We have outlined how both practices and assumptions are often in fact inappropriate to the contemporary contexts in which they are employed. In seeking now to open up some possibilities for realigning practices and assumptions about the future, we change our focus to the conceptual frameworks that support these assumptions and allow them to function. We therefore move from the sociological level of practice to the philosophical level of the conceptual structure of these practices. As detailed in previous chapters, the ways futures are constructed in contemporary industrialised societies reflect a long evolution, through which the telling and taming of embedded individual and social futures has gradually been replaced by the ceaseless transformation and trading of empty futures*, made possible by the conceptual tools provided by mechanistic science. To understand the future as an abstract*, disembedded* realm, accessible through methods of prediction, is to understand it as a territory belonging essentially to no-one and hence one that is open to seizure by anyone in the present. It is to relate to it as a present future* and as a measure of exchange value. However, as we have argued, the practices we have analysed do not help us to comprehend the futures they unleash.

This is because the way we tend to decontextualise and empty the future ignores the latent futures* hidden in networks of processual interdependencies that they set in motion. They create new uncertainties which return us to the problem of how to tame the future, which has been displaced by that of how to transform it. To tame the future is to relate to a future we do not own, rather than to frame it as a storehouse of potential, limited only by the images of present futures we can create. It implies recognition of the responsibilities that are attendant on futures-creating action. But, as shown in Futures Traded, to empty the future by viewing it solely in terms of economic values such as profit and growth, and for the purpose of maximising short-term benefit, fragments and destabilises it. This fragmented future makes impossible such communal practices of future-taming as the *Kula* that institutionalise responsibility for the future. As shown in Futures Traversed, current technologies intensify this difficulty by foreshortening our horizons and further collapsing futures into the present through the valorisation of speed*.

In order to socially acknowledge once again a responsibility to the future presents of others, we need to re-embed our understanding of the future within the networks of temporal interdependencies into which the timeprint* of our practices inserts us. We therefore need to develop a different conceptual basis for our relationships with the future, one that re-sensitizes us to the kind of perspective that informed traditional practices of telling and taming. It is this task that we will attempt in this and the next chapter, offering *a new set of conceptual coordinates* for mapping processes of futures-construction. The crux of the problem is our social tendency to actively disconnect the future from the present. This pushes us to view it *through* the present as an abstract possibility which may or may not emerge. This tendency affects all the conceptual resources we can bring to bear on mapping our futures. Decontextualised, emptied, and open for transformation—the present future is a future to which we no longer feel any intrinsic connection. Whatever it will contain, we feel that it is subordinate to what happens here in the present. The more we live for our now, the less we can connect with the 'nows' of others yet to be born, the future presents of generations to come.

To begin to alter our relationship with the future, the concepts and metaphors through which we approach it must be changed (Lakoff and Johnson 2003/1980). We must draw into our thinking, imagining and feeling the 'shadow side' of futures as latent processes on their way to emergence. Instead of conceiving of futures simply as the products of our actions and activities in the present, we have to understand the futures societies create as swelling up within them, always on the way to unfolding. The future in this sense is not abstract, not empty, and not simply open to transformation, but is instead *living** within the present. It inhabits the relations that establish the interdependence of things, and which contain the potential for producing unintended and unforeseen consequences. If our central metaphors for the future were to become ones that depicted it as a ceaselessly unfolding and refolding life, like Baruch Spinoza's (1632–1677) *natura naturans**, 'nature naturing', then it might be possible to change our future orientation (Spinoza 1992/1677). Perhaps we would no longer empty the future in order to collapse it faster and faster into the present. It might be possible once more to build a relationship with future presents into action, knowledge and ethics.

To change the conceptual basis of knowledge practices, we can draw on and develop resources from the history of Western philosophy that give us access to a different image of the future. It is true that over two thousand years ago, Plato gave us an image of ultimate reality as an eternal present, a single moment within the mind of God that contains all possibilities within it, with time (including past and future) itself being illusory. Nonetheless, the Western tradition also contains other, radically different ideas about the nature of time and, more importantly, about the future and its relation to the present. Thinkers from G.W.F. Hegel to Henri Bergson, Alfred Schutz and Jacques Derrida have shown that the present as such is more a construct than a basic reality. In this chapter, we will draw on the work of four thinkers to develop two new concepts which we shall call the *lived* future* and the *living* future. Together, these concepts will enable us to unsettle some of the future metaphors that govern contemporary knowledge practices, by placing the long-term, latent future at the heart of action, knowledge and ethics. For these thinkers, the future is neither an empty space, and nor is it just something which is imagined or thought about. It is a latent but real feature of being as such.

The lived future refers to the future experienced as a constitutive element of the present, something without which there could not be a temporal aspect to experience. A key thinker in this regard is Martin Heidegger (1988/1927), who offers an analysis of human being which locates its distinctiveness in its future-orientation. To be human is to be ceaselessly becoming, constantly beyond ourselves, understanding the present and past by projecting ourselves into potential futures. No consciousness of the present is possible except from within this horizon, in which we become aware that objects and people make up a structured world that matters and is meaningful to us. Our awareness that we will die places upon us a unique responsibility for the meaning of our lives and the care of our projects. The future in this sense is therefore neither empty nor abstract, but is a *lived* aspect of our experience that embeds us within a meaningful world.

For Hans Jonas (1982/1966), the second philosopher whose inspirational work we would like to draw on here, future-orientation is a quality that unites us with nature, rather than separating us from it. To be a living being is to strive forward into the future. All living creatures have interests, in the light of which aspects of their environment are revealed as salient to them, and which both grant to them a sense of bodily orientation in space and shape rhythms of activity through which these interests are pursued. Further, all living beings possess what Jonas calls *sense*: they might not be self-conscious, but they are 'aware' of their environment to varying degrees. All can perceive significant differences in it and change their own state in response to anticipated futures. When self-extension into the future comes to be seen as part of all organic life, then the difference between humans and even unicellular organisms becomes one of degree, not of kind. The rhythms that govern each level of organisation of an organism all contain anticipations of the future.

This continuity could even be extended to inorganic matter, in so far as it is understood as self-organising, as for example in the spontaneous self-assembly of chemical structures. Gilles Deleuze (1994/1968) and Deleuze and Félix Guattari's (1988/1981) work depicts a world in which the extension of beings into their futures is a feature of matter itself. All stability is seen as a local effect of wider patterns of instability and divergence, of a world in which matter is primarily fluid and energetically unstable, and only secondarily solid. The conservation of energy within an organic or inorganic system, an animal or a crystal, is the condition of its self-preservation, and thus of its possession of a continuing present. But the existence of such a system is dependent upon transfers of energy into and across it from other systems, all of which are thus interconnected in symbiosis, creating wider ripples of change and transformation. Complex symbiosis gives to everything both an actual and a virtual dimension. In such a world, stable systems are always open to unpredictable and radical transformation, as organised matter always contains within it futurity, that is, the active potential for further transformation. This opens up the *living* future, the future as already active in unforeseeable ways within the present.

These thinkers of lived and living futures thus return us to a key insight in Aristotle, which has been largely excluded from knowledge practices such as mechanistic science and economics. Aristotle proposed that we cannot explain how things are the way they are just by relating their current state to a mechanical 'push from the past'. 'Cause' is generally used to translate his concept of *aition*, an inherently ambiguous term which means whatever one can cite in answer to a 'why?'

question. Aristotle recognised that explanations for the emergence of a phenomenon can be given from various perspectives, of which he identified four: a formal, material, efficient and final *aition*. Of these, only that of efficient *aition* comes close to the mechanistic notion of cause, according to which the future can be read off from what we know of the past. Of the others, the idea of a final cause, a *telos* or purpose, refers to a future state which is encoded in the present of a system. After the rise of mechanistic science as described in Futures Traded, the idea of final causes began to be decisively rejected as unscientific. However, although the thinkers we examine in this chapter have all rejected any deterministic overtones of Aristotle's concept of final cause, they have resurrected the idea that the future dimension of things cannot simply be left out of any attempt to explain their present. Indeed, they propose that the different ways in which all kinds of beings are never 'finished', never quite self-identical, and always somehow 'beyond' themselves are an essential part of understanding what they are.

Futurity as Care

For Heidegger, the future is real because it is a constitutive element of all aspects of being human. In his early work *Being and Time* (1998/1927), he analyses the temporal structure of human existence in order to characterise what it is to *be* human, defining it as a particular mode of being he calls *Dasein** ('being-there'). Human beings are, Heidegger proposes, the kind of being whose experience is always of a *world*—a significant whole into which they are inserted at birth, and within which their lives take on meaning. Our primary day-to-day concern is our immediate interests, and it is on this basis that we assess the significance of the objects and people around us. Concentrating on the short-term, we tend to equate reality with what we see and hear in our immediate vicinity. Confidence in our immediate perceptual access to things places us in the position of an observer who is able to 'step out' of the continuous flow of time, extracting events from this flux, and placing them in sequential order. From this privileged position, we are able to understand the course of things and to intervene in it when appropriate. This is comparable with the position of what the ancient Greeks called a *theoros*—one who is capable of observing, of "pure beholding" without directly participating (Gadamer 1994/1960: 124).

From this point of view, writes Heidegger, "Being is that which shows itself in the pure perception which belongs to beholding, and only by such seeing does Being get discovered" (Heidegger 1998/1927: 215).

For Heidegger, the way we experience the world turns on how it is *disclosed* to us. What this means is that we do not observe the world from an idealised point outside it, from where it is possible to objectively reconstruct the course of events. Rather, 'disclosure' signifies that the world is always only revealed to us in relation to what appears possible for us to accomplish (Heidegger 1998/1927: 33). To be human is always to be concerned with one's "to-be [*zu Sein*]" (Heidegger 1998/1927: 67), and so knowledge and action are always inseparable.

This means that we understand what we encounter within a halo of our own projected future potentiality. The demands of this foreshadowed future spur us to action, while the past is a set of resources that enables action in the present in the light of these demands. Consequently our responses commit us to a more or less definite 'range' of futures. The role of the future in action means that "[t]he primary phenomenon of primordial and authentic temporality is the future" (Heidegger 1998/1927: 378). Our primary relationship with the world is thus *constitutive*: we do not passively take in, moment by moment, the world as it is objectively. The world *matters* to us because it is the *mise-en-scène* where our possibilities will be played out, and consequently we are active in forming the meaning of the things and people in it.

Human being is *Da-sein* because all human life is lived in some 'there' (*Da-*) or other: a viewpoint from within which the world is revealed and becomes meaningful. My 'there' is not yours, and vice versa. This is because every human being is not just 'ahead of itself', but is what Heidegger calls a "thrown possibility" (Heidegger 1998/1927: 183). Our 'there' is always characterised by 'thrownness [*Geworfenheit*]', a sense of being cast into the midst of a world that is already loaded with the interpretations and meanings into which the possibilities of others have coalesced before we were born. Therefore our past and present have at one and the same time three aspects: they are the *congealed futures* of our predecessors, are pregnant with their unrealised potential, and are carriers of futures begun by them which are still working themselves out. Our past and present are therefore both *products* of former futures, and *processes** of still-latent ones. Similarly, future generations will find themselves thrown into a world created through our actions. The 'there' of human beings can therefore never really coincide with the viewpoint of the *theoros*. No-one can ever be outside of the active interweaving of the past, present and future of a collectivity within which the significance of the world is experienced.

So the challenge human beings face is to understand the meaning of a world made by others. This depends on the totality of the ways we relate ourselves to the potential meaning of the things we encounter—emotionally, imaginatively, through memory, reasoning, and the variety of ways we have of interpreting contexts. Heidegger sees our 'there' as weaving experience from a variety of ways in which we involve ourselves with things (Heidegger 1998/1927: 177, 182). We are thrown into a world that comes loaded with a natural and social past that we cannot get behind, but which concerns us. It concerns us because we understand the significance of the things we encounter in relation to our own futures. Involvement with things at every level of our being means that our most human feature is that we *care** about the world. In the widest sense possible, we are concerned for what will and might 'become of things' in the future. Everything we do is shaped by this basic orientation.

So for Heidegger, human beings do not simply passively receive information about the state of things 'out there'. Nor for him is the future simply empty, purely open, or mathematically abstract, and hence indeterminate. Instead, it is always experienced as having a singular content. Granted, this content is not determinate or finished: it is not a *product*. It is however, *determinable* (Johnson 1921: ch. 11)—it is like material that, thanks to its inherent structure, can be formed into certain shapes and not others. The extent and precise character of this determinability derives from a unique 'there', the singular perspective of an individual on their own future. This real future is therefore a dimension of our existence in the continuously passing present, and so is a future that we *live*. Being constantly beyond ourselves, we strive to understand our potential so that we can understand what we are becoming in the present. We 'stand out' within time, projecting ourselves into the future, which we know will be incomplete, for part of our future horizon is our understanding of ourselves as mortal. To make a very Heideggerean pun, we 'stand out' within time because there is always something *still outstanding* in our lives, something which has yet to be settled—our own death (Heidegger 1998/1927: 179). What makes us beings that live (in and through) the future is the fact that we can never be *complete*, in the sense of realising all our possibilities. Whatever we become, achieve and produce, our 'ownmost possibility', our impending death, cannot be realised without bringing our becoming as *Dasein* to an end. This is why, for Heidegger, we can never occupy the standpoint of eternity, or even understand what eternity means. It is because the idea of the eternal present that belongs to God alone is one of completion: God is unchanging and perfect. What it would be like to experience a moment that is truly present because its meaning is *complete* is beyond us. Nonetheless, whenever we adopt the perspective of a non-participatory observer, we are seeking a simulated version of eternity in which we are separated from the world of change and becoming.

In Heidegger's work, death is therefore the ultimate horizon within which the world takes on significance, the "possibility of the absolute impossibility of *Dasein*" (Heidegger 1998/1927: 294). From this point of view, a mother's understanding of her daughter's potential is rooted in her knowledge that the significance of childbirth, and of the life of the child, derives from the child outliving her parents. What it is like to be a parent pulsates with this understanding of the future. Rather than being experienceable events, birth and death are the roots of the mother's way of being and define the limit of her possibilities. For Heidegger, then, we always interpret the present from within a standpoint that has a horizon of futurity, laid out by our multi-level understanding (emotional, imaginative, cognitive...) of what we might become. The reality of this future is the light and shadow simultaneously cast by it on the present, changing the way the world appears to us. This *lived future* is not therefore objectively *known* or predicted: our viewpoint on the world does not, and cannot, present the future to us in this fashion. Nonetheless, the insistent demands of future potential and present latency are accessed through the modes of perception and attunement which go to make up the existential bedrock of the attitude that Heidegger calls *care*. To tune in to my or our potential requires more than calculation of empty or abstract futures. It demands an informed, imaginative and integrated sensitivity to the shifting dimensions of potential that fringe the present. This sensitivity makes us creatures that are intrinsically incomplete, and who always possess a sense of 'something still to be done'. This is the core of being human, an experience of a living, continuous present composed of ceaseless change that always points beyond itself, towards the horizon of what it is becoming.

Anticipatory Bodies

Humans are also part of nature, however. The continuity between present and future Heidegger describes also reaches deeper into us, beneath the conscious level of experience, and includes a kind of biological experience of the future which we share, to varying degrees, with other organisms. This is a lived future rooted in the deepest level of our bodily existence, the metabolic exchanges of energy with our environment that govern our physical state. The philosopher Hans Jonas has given an account of this biological experience of futures, in which organisms (including human beings) maintain themselves by sustaining conditions that are favourable to them through their powers of adaptation and anticipation. In Jonas' view, all organisms share with humans a sense that the world *matters* to them. This is because all organisms have interests related to the ways in which things can go well or badly for them (Jonas 1982/1968: 84–5, 126). If they have interests, then they have futurity, as they strive after what furthers their interests and avoid what is detrimental. All organisms possess a kind of 'sense' or power of consciousness by virtue of the specific sensory and motive means by which they interact with their surroundings. They are able to perceive differences in their environment, distinguish beneficial things from bad, and orient themselves within their environment towards these. Their interests therefore make certain aspects of their environment perceptible and *salient* to them. In this sense, all subjectivity, not just that of humans, is an active 'interpretation' of the environment rather than a mere passive receiving of bits of data.

Just like humans, organisms are not only part of an environment, as they also possess a *point of view* on it together with a *project*. In relation to this project and viewpoint, the organism's environment has a kind of temporal boundedness to it. In a similar way to how humans experience their lives as having a unity in time, determined by their concern for how their projects will turn out in the future, organisms remember pasts and anticipate futures in the course of pursuing their interests. It is from within this bounded perspective on the world that an organism's surroundings take on salient and non-salient aspects. Thanks to its remembering and anticipation, the environment is *meaningful* for it, rather than being a mere collection of physical bits and pieces. Heidegger describes meaningfulness as dependent on care, a sense that things matter. Jonas sees, at the biological level, a similar 'attitude' pervading the organism's interactions with its environment—a sense that there are things that must be done, and therefore that certain aspects of the world are 'salient matters'. This basic self-concern is what he calls *conatus**, or striving to persist in being (Jonas 1984/1976: 72–3). The future of an organism, from its own 'point of view', is therefore neither an empty future nor the abstract future of natural science. Rather, it is *lived* in a similar sense to the future of Heidegger's *Dasein*. It is an implicit structure which 'pulls' the organism towards it, shaping its actions and progressively unfurling its potential along the way as it strives to realise itself.

It is true that an organism is not conscious of its future in the way that human beings can be conscious of their lived futures. Its unconscious nature nevertheless enables organisms to adapt to novelty by projecting forward their needs and anticipating potential obstacles. The organism's metabolic system and its various subsystems have their own temporal rhythms. This allows the organism to anticipate the need for food, sleep and so on, together with dangers that might accompany any particular way of satisfying its needs. If the basic temporal structure of an organism is projective in this way, then there can be no organic life without continuity between its latent and potential futures on the one hand, and its living present on the other. In the same way that *Dasein*'s futures are the background against which the present becomes meaningful, the anticipated future of the organism is the background of salience against which adaptation and action are possible in the present.

Beneath the temporal character of the life of *Dasein*, there is another experience of futures which it shares with other organisms to different degrees. This is its *intensive* awareness of its living present, into and out of which the dynamics of evaluation and action extend. It is coloured by variations in the cyclical rhythms that constitute bodily processes, but is always focused on how these rhythms can be maintained in harmony with each other. Thus Heidegger's presentation of human experience as a kind of total involvement with things is complemented by Jonas' idea that a body's metabolic system itself projects and orients itself towards particular latent or potential futures. Jonas therefore describes another way in which we live the future, one that reaches far beyond the level of Heidegger's description down to the rhythmic oscillations of biological systems. This relates us both to other, deeper temporal levels of our *organic* being, and to ones deeper still which belong to our *inorganic* being. Futures are experienced by organisms, including humans, as an effect of their own needful engagement with the world. But there are wider patterns of potentiality and latency that make possible these forms of engagement, which we have identified as *living* futures. These are what we shall examine next. They embrace all levels of reality from the inorganic upward, including the dimensions of lived futures identified by Heidegger and Jonas.

Extension without Boundaries

Nature is both organic and inorganic. We can imagine the continuity that links a mother and her daughter with their pet gerbil, with the runner beans in the vegetable plot and the cypress trees in the neighbouring park, with the insects skating on the pond around which they grow, and even with algae blooming in the water. The attribution of some degree of sense to all these organisms might not overstretch us. But is there anything that links what it is to be human with the silt at the bottom of the pond, or the rotting vegetation that mixed with it, other than certain chemical elements? Can we meaningfully say that a chemical element possesses a 'viewpoint' on the world, in which future-orientation enables it to 'interpret' and 'experience' its environment?

This idea is not unknown to contemporary biology and chemistry, into which the development of complexity theory has introduced a conception of explanation which does not fit the traditional mechanistic model. Basic natural processes can in certain conditions exhibit tendencies of change that are imbued with a horizon of futurity, through which they demonstrate what Ilya Prigogine and Isabelle Stengers (1984: 14) have called a "prebiological adaptation mechanism", through which they become "able to perceive, to 'take into account', in [their] way of functioning, differences in the external world (such as weak gravitational or electrical fields)".

This concept of matter takes us close to Jonas, who suggests we should extend the idea of striving (*conatus*), and therefore sense, to inorganic matter, which is, like living organisms, always beyond itself, already heavy with potential:

> Admittedly a 'psychic' aspect always adheres to striving as such. And why not? 'Psyche' and 'selfhood' are not identical, and the first may, in a generalised form be an appurtenance of all matter, or of all material aggregates of certain forms of order, long before it attains individualisation [...] (Jonas 1984/1976: 72–3)

To extend the idea of continuous becoming, and with it, lived futures, 'all the way down' to inorganic matter is a goal of the work of Gilles

Deleuze and Félix Guattari. They explore how matter can organise itself into systems with regularised temporal structures of the kind described by Jonas. To complement the idea of an inorganic lived future, we now want to develop from their work the concept of *living futures*, that is, latent flows of potential which, under specific conditions, congeal into organised physical structures with lived futures, such as organisms. To think of matter as possessing living futures allows us to view it as having destinies—'ranges' of potential futures—which play themselves out in inherently unpredictable ways, across many different temporal scales.

Some examples of how living futures are played out and congeal into actual systems would be evolutionary processes such as genetic drift and symbiosis, and social processes such as migration and technological innovation. For Deleuze and Guattari, the transition from potentiality to actuality often occurs through the novel symbiotic combination and transformation of existing inorganic, organic and human/social systems, in ways which may turn out to be beneficial or harmful. For example, symbiotic evolution of living systems across thousands of generations tends to benefit the organisms involved—e.g., some species of orchids have evolved to mimic female wasps in order to secure the pollinating services of male wasps. By contrast, the introduction of artificial radioactive material into the environment leads to long-term harmful transformations of organisms, including humans, which will also unfold across thousands of generations. These examples show that *complex processes of mutual forming and shaping operate across the inorganic, organic and human strata of the world.*

This idea of mutual forming is vital for our attempt to build concepts of the latent future into our metaphorical and conceptual vocabulary. To understand more of what it implies, let us consider two strongly contrasting ways of participating in the world described by Deleuze and Guattari. These they refer to as an *architect*'s 'way of being', and that of an *artisan*. In order to understand these modes of participation, let us imagine an example, a little girl who constructs a tower of building blocks. We can picture her—as, indeed she might picture herself—as the architect of her project, beginning with a mental plan before surveying her materials, a pile of bricks which, in comparison to her imagined tower, lacks any order. Rummaging through the bricks in search of the ones she requires, she begins fitting them together. Her idea of what her tower will look like might appear to her as somehow more real than the pile of bricks in front of her. What they lack is the *form* which she strives to give to them. They seem to be a passive material that receives order from an external source, i.e. her mental plan. Her relationship with the bricks is hierarchical, or more specifically, *architectural*. She might imagine herself as the architect of the final product, and think of the future as being gradually transformed into one in which her plan becomes reality. The future of the bricks, as far as she is concerned, is a present future, one which contains a number of possibilities between which she chooses based on her desires. On this basis, we can compare knowledge practices of future transformation and colonisation that construct present futures with this architectural model.

For Deleuze and Guattari, this architectural image of order is an ancient and enduring feature of Western thought and culture. For example, it appears in one of the earliest recorded philosophical (as opposed to religious or mythological) stories of creation in Plato's *Timaeus* (Deleuze and Guattari 1988/1981: 369). Plato's narrative tells how the universe was created by a god from unformed matter, using an idea of a perfect universe as a model for his desired future. The method is architectural and mathematical, proceeding through measuring and division to turn an idealised blueprint into reality. The aim is to create an image of solidity and permanence in a material which is unstable and constantly changing. In other texts, Plato describes the role of an architect in very similar terms. The architect must possess the correct *techne*, a set of rules that can be set down and taught, and which consists largely of knowing how to *command* matter, using a measured process of division and allocation. Plato explicitly distinguishes this kind of skill from the 'knack' that a skilled carpenter or stonemason has of forming a shape from material by responding to and cooperating with its internal structure.[1] As this skill results from the prolonged training of the artisan's body rather than their intellect, and has a method that cannot, unlike a *techne*, be precisely formulated, it is denigrated by Plato. For Plato, the artisan is a passive source of labour-power whose activity must be informed by the architect's commands, since the real power behind any creative process is an intellectual idea of the future. However, Deleuze and Guattari suggest that the assumption that we are like architects, as opposed to labourers who simply do what they are told, separates us from the futurity immanent in all matter.

Let us follow Deleuze and Guattari and shift our perspective to imagine a process of construction from the standpoint of the artisan

[1] On this distinction, see the illuminating discussion by John Protevi (2001: 122 ff).

(Deleuze and Guattari 1988/1981: 363), who does not try to impose form from outside. Instead of viewing matter as simple 'stuff' that must be formed, artisans (like metallurgists, sculptors and woodcarvers) appreciate that it has an inner structure which, in relation to their practical skills, already foreshadows a set of futures, just as a lived human or organic future does. To coax forth from this real potential an actual form, the artisan must *inhabit* the material (Deleuze 1994/1968: 36–7). He strives to follow the salient traits it bears within it (such as knots of wood, twists of fibres, or the striations within marble) to which the intensive training he has undergone has made him sensitive (Deleuze and Guattari 1988/1981: 380–2). The process of forming, therefore, is two-sided, shaped both by the limits of the artisan's capacity for total involvement in the material and the precise nature and direction of the material's resistance to his involvement.

The evolution of a material form is therefore an ongoing interplay of forces deployed against each other from within systems that are already organised (socially, psychologically, organically, and/or chemically) in some way. It is not determined by an idea that exists before the process begins, but is a process which unfolds by way of contingent 'negotiations' between these shaping forces. The potential of each for producing forms is only revealed when they come together: in other words, when they combine by interacting in some way, the potential and therefore the futures of each, are augmented in perhaps surprising ways. Deleuze and Guattari's general argument is that this kind of mutual, symbiotic augmentation of potential is a useful metaphor for understanding the emergence of novelty as such. We want to follow them here by suggesting that, rather than being restricted to human 'artisans', this capacity is common to varying degrees to all matter, and is expressed in ways which depend on the unique modes of interaction that are possible between specific entities. Further, as we noted above, these interactions can cut across different 'levels of reality', perhaps bringing together groups of humans and/or social phenomena (such as technologies) with organic and/or inorganic systems. Their unpredictable results can be either beneficial or harmful to the individuals and/or groups involved.

Based on their respective forms of organisation (social, psychological, biological, and/or chemical), systems possess specific capacities for 'projection'. These capacities enable even the most heterogeneous entities (such as a human being and a block of marble) to 'fit together' in unforeseen ways which produce changes in both. Through these entirely physical processes, new material forms and systems of meaning can be generated. For Deleuze and Guattari, there is therefore a second physical *process* dimension to things which accompanies that side of them which is a physical *product*. The product side is composed of material components interacting within systemic relationships that are governed by individual and cross-cutting rhythms. But these systems of *actual* components also have the aforementioned process or *virtual* dimension. They can project themselves forward into the future by anticipating and adapting to circumstances, as Jonas suggests. Their capacity for future-directed transformation however extends deeper and wider to the level of what we have called their living futures. For example, processes of evolution are not salient aspects of the environment for an individual organism, as they are not perceptible to it. Nonetheless, all individual organisms are expressions of these processes, products of the genetic and environmental factors that condition the metabolic systems that make them what they are. In other words, the *lived futures of material systems*—the structures of chemical elements, the way that organisms adapt to their environments, the way that humans care about their lives as 'projects'—*extend themselves into living futures that combine within them many different lived futures.*

Thus, we can say that what Deleuze and Guattari describe as the virtual dimension of matter comprises in it both lived *and* living futures. Organised matter, at all levels of complexity, from the purely chemical to the psychological and social, tames or 'captures' potential or future-orientation for the sake of prolonging the stability of its own form. To adapt to circumstances, it has to extend itself in time beyond the oscillating rhythms of tension and relaxation that govern those parts of itself which are actual and 'finished'. In doing so it develops certain capacities, and as a result, it also attracts to itself untamed potential. These capacities of a system mark out a larger or smaller determinable range of futures for it. The fates of these futures are various. Some may remain entirely unrealised whilst others gradually unfold. Alternatively, the whole range of available futures might be radically transformed by an unexpected encounter with another entity or entities. For Deleuze and Guattari, this virtual aspect of things, both tamed and untamed, is no less real than their actual structure: it surrounds them like a halo. When two or more entities manage to form a relationship due to a certain 'fit' between their living futures, it is their respective 'haloes' of potential for transformation that overlap, merge and then become expressed through this relationship. The relationship is the catalyst for their transformation, and actualises their potential in ways that are not available to them on their own, and which will alter both their own lived futures, and wider patterns of change which extend beyond them.

Because the virtual dimension of things includes, but cannot be reduced to the portion of it which is captured within the lived futures of individuals, it is without any 'centre' as such. Because it has no centre, it comprises more than just lived futures, which always depend on a perspective, a locus of experience. For example, symbiotic co-evolution of the kind we mentioned above in relation to the orchid and wasp emerges from local encounters between individuals, which in turn engender processes of transformation which stretch out over millennia. Nonetheless, these long-term processes are themselves expressions of the potential contained in organisms, understood as highly complex combinations of ordered catalytic chemical reactions that last for fractions of a second. Suspended between the long-term living future and the lived futures of chemical reactions, individual members of a species project their own lived futures by systemically regulating and rhythmically coordinating billions of chemical reactions, and thus live out their allotted span. In the process, each adds the potential contained within its own life to the living future of evolutionary transformation. Because of the massive diversity of the manifold processes of transformation within which individual entities are implicated and which surround them, living futures of many different scales are always invisibly under way in and through these individuals, and thus subject to unpredictable changes of direction. The unforeseeable ways in which living futures influence and become expressed in actual physical transformations can be beneficial or harmful for the individuals caught up in and buffeted by them.

We will now extend Deleuze and Guattari's concepts still further, and understand how either malign or benign outcomes are produced by symbiotic mutual forming. To do this, let us consider the differences between instances of natural forms interacting, and through their interactions discovering a new symbiotic relationship, and cases where technological intervention creates a catalyst for new relationships of this kind. One example of the former might be the relationship between tube worms and symbiotic bacteria that live at hydrothermal vents under the ocean. Over millennia, these worms have lost their entire digestive tracts, and rely wholly for nutrition on bacteria that live within them. For their part, the bacteria metabolize chemicals which the worms supply to them. Another process of symbiosis is evident in 'entrainment', such as when an animal adapts its rhythms of sleeping and waking to changes in the length of day and night throughout the year. Here, the two systems in interaction are a planet and its moon orbiting a sun, and a single organism on the surface of the planet. When two or more entities come together in this way, a complementary relationship emerges between them based on their respective capacities for participating in processes of exchange. For example, the metabolism of the worm and that of a particular bacterium are capable of adapting to each other. There is a certain 'fit' or harmony between the respective ways in which they extend themselves temporally to seek out and metabolise what is salient for their survival. Although they are heterogeneous systems, they are able to reach, over time, a certain 'accommodation' due to this compatibility between their lived and living futures. They embed themselves gradually within an entirely new and hitherto unforeseen relationship, and a new living future.

By contrast, the examples of the social deployment of technology we have discussed in previous chapters furnish us with ample illustrations of how the transformation of living futures can result in long-term harm. The heterogeneous combinations into which technologies can lead us have, as we have previously emphasised, a pronounced disembeddedness. It is this quality that marks these combinations as different from the processes of co-evolution that occur between heterogeneous individuals in nature. Here, co-evolution occurs over long timescales, leading to gradual entrainment and the production of a common, and most often beneficial, living future. This is not the case with the combinations which technologies produce. Often, we have argued, technologies are employed as means of abstracting processes from context and altering their duration, mirroring the broader knowledge practices analysed in Futures Traded which focus only on short-term benefits and the valorisation of speed. As we have proposed in Futures Traversed, they combine different elements—inorganic, organic, social—with different potentials for transformation in a relatively short span of time, perhaps even in a single moment of transplantation. Through these acts of combination, they unleash new processes of transformation whose timeprint is vast in extent. They can act as catalysts of uncontrolled change.

The result can be unexpected and radical short-term shifts in the ecology of an organism or organisms to which they cannot adapt, and which change the living future of other creatures distributed along a much wider network, possibly encompassing the whole planet. For example, consider Colborn's research on the widespread bioaccumulation of synthetic chemicals found in plastics (Colborn et al. 1996). From this form of contamination, in which pollutants were introduced into the environment over the course of a few decades, came the bioaccumulation over the same period of pollutants in individual animals and ecosystems spreading right up the food chain and across the globe. The projected consequences, such as reduced fertility and birth defects, threaten a new living future which did not exist before this technological intervention, and which possesses a potentially vast timeprint stretching ahead of us. Here, it is the very stability of the plastics that is the problem. The qualities of chemical inertness and durability for which they are prized are ones which appear to mark them out as having little potential for producing unforeseen side-effects. They appear to be the perfect disembedded technological solution, a product that no longer contains latent processes that could overturn their intended purpose: as they are chemically inert, they can be used in all sorts of roles in a huge variety of different environments. However, their inertness gives them an unintended capacity to create disruption. It means that they persist in the environment in such a way as to facilitate their ingestion by all manner of creatures, which results in a gradual accumulation of chemicals all the way up the food chain that can disrupt the functioning of animals' endocrine systems. The result is that the living future the ingestion of chemicals produces is one that interrupts the lived and living futures of organisms, their capacity to reach beyond themselves, to reproduce and to evolve. The apparently inert product still carries with it a disrupting potential, but one that remains invisible to the decontextualised perspective that created it.

Reflections

By changing our metaphors, and thereby understanding futures once again as embedded, as lived and living, we are able to see the roots of the unacknowledged problem that contemporary knowledge practices create. Heidegger, Jonas and Deleuze and Guattari all suggest that the future is more than the empty, transformable, colonisable space that is all too often constructed by contemporary knowledge practices. They propose instead that, in our continuous, changing present, the future is already real and active, but in a specific way. For Heidegger, humans are, in the very roots of their existence, always beyond themselves. We 'stand out' into the future, and interpret our present through the multiple ways in which we anticipate what we might become. Jonas extends this sense of futurity further. An organism is inherently adaptive and anticipatory, improvising responses to its environment. For Deleuze and Guattari, finally, any material system can produce new forms in unexpected ways, and to varying degrees. The scope of this potential development is of unforeseen and unknowable extent. Just what and who a little girl might become is by no means settled by the fact that she has learnt to build a tower of bricks, but the future processes of becoming into which she will move have already been prepared. The future as virtual, living future is therefore not 'virtual reality' in the sense of an unreal simulation of what is real: it is, instead, the latent yet material dimension of that which already exists, and which is always at work, creating patterns for near and unimaginably distant futures. As this virtuality evolves into new, actual forms it may eventually be lived by and take on significance for those individual 'components' that share in it. When it is lived, once it becomes incorporated in bodies and in the social meanings by which humans project and organise their lives, it may emerge as beneficial or harmful.

As we have seen in this chapter, the belief that the future is empty and colonisable should be seen as a perspective error. It is one which is repeated, intensified and driven ever deeper into the foundations of social practice by a whole host of knowledge practices (such as the use of mechanistic science in politics, economics and technological innovation). It ignores other, more primordial modes of futures-construction, some of which are directly lived at the level of our social and or biological being. Others are rooted in connections between our social practices, our bodies and the ecosystems which we inhabit. These living futures spread out around us, embedding us in deeper patterns of change with rhythms and durations that reach far beyond our lives. In contrast, the long-term effects of high technology arise from scientific innovations whose potential cuts across several levels of reality, connecting social practice with the microstructure of nature. The capacity of these technologies for producing unintended consequences is therefore proportional to the extent to which they forge connections between so many different futures—social and individual, cellular, ecological, chemical, radiological. They consequently unleash new virtual potentials that could not have been foreseen from the 'architectural' standpoint of the present future, which is constantly employed in boardrooms, laboratories, workshops and factories where futures are planned and decided, based on knowledge of the past.

If the future is real and always already with us, but in virtual, anticipated and projective form, then our relationship to it is misunderstood if we think of ourselves as colonisers of an empty future ready to be occupied. We are not related to living futures in the same way as inorganic catalysts or animals. Our level of lived involvement in them is characterised not simply by unpredictable becoming, or adaptation, but by projective care for the meaning of our own lives. If the future is virtual, projective and *therefore* real then we are its artisans rather than its architects, and the relationship between an artisan and her material is also one of care. Carving the wood entails being sensitive to the traits of the wood and to their inherent future presents, rather than ignoring them in attempting to realise a blueprint. It is only in this way that carving is possible. The woodworker is not the architect of the figure she carves, and nor is the mother the architect of her daughter. Their relation to the future of the objects of their concern is one of care, in which they seek, through their sensitivity for what might emerge, to accompany the *desired* virtual potential of the living present to its full realisation in an awaited future present, and perhaps beyond. In the next chapter, we will take up the relation between our actions and their consequences once again, this time from an ethical perspective. We will seek to make a connection between our involvement in and care for the futures that we live, the living futures in which they are implicated, and our ethical relation to both.

CHAPTER EIGHT

FUTURES TENDED

Introduction

By opening up an alternative philosophical perspective on the future, we have laid the foundations for a different understanding of the ethical content of the knowledge practices* through which futures are constructed. In this chapter, we develop further the reflections of the previous chapter concerning the depth of our immersion in lived and living futures*, the ways in which the future *matters* to us. In doing so, we will see how, when social practices produce empty futures*, their resulting focus on the interests of the present makes it difficult to provide an ethical context for social practice and policy that is appropriate to its timeprint*. One way to respond to this lack of context is to provide a *new* context, using the concepts of lived and living futures to reinterpret the meaning of responsibility and obligation, and to mobilise ways of thinking about responsible action that draw on our social memory of futures*. By allowing us to imagine different ways of acting responsibly in creating futures, these ideas will provide us with some new conceptual coordinates for thinking about the ethical underpinnings for our relationship with the future, and for reshaping the legal and thereby the political expressions of our responsibilities to it. They will help to restore a sense that the future matters.

Responsibility for the Present

The exact social meaning of responsibility varies from situation to situation, and indeed can vary depending on which aspects of a single situation are being considered. Most often, it is understood within a legal framework in which it is interpreted as *liability*, and/or a moral one in which it is understood in terms of *blame*. For example, a car driver who is involved in an accident in which someone else is injured can be *held* responsible for the injury after the fact. She could *assume* responsibility for it by reporting her part in the crash. All the facts could point to her *being* responsible for what has happened.[1] All these ways of attributing responsibility have in common certain key features. It is because of these features that they can become components of legal, ethical and policy practices that interpret liability and blame in ways which reinforce and amplify the harm done by emptying the future and making it unreal.

These key aspects concern the way knowledge of causality is necessary for any understanding of responsibility. Once the crash has occurred and an injury has been caused, then each aspect of the driver's responsibility suggests that a specific causal relationship exists between a past moment and the present one in which her responsibility is established, claimed, ascribed or whatever. If she is held responsible, then whoever accuses her must do so on the basis of evidence that her actions led to the injury being caused. This evidence therefore establishes a timeline of events, a sequence of facts that together *make* her responsible. To own up to being responsible is just to admit that such a sequence of events exists, and that they could not have happened without our actions. There may be a series of these actions, or there may just be one decisive action. Responsibility therefore rests on evidence of causation. What makes the driver responsible is a series of facts which point to her decisive role in causing the accident. By establishing a timeline between one point in the past and the moment of the accident, knowledge practices (such as forensic science) allow causation to be established.

There is another related meaning of responsibility, however, one which will be crucial for our examination here of the failings of contemporary concepts of responsibility with respect to the future. The driver may claim that she *acted* responsibly, with due care to the likely consequences of her actions, but things just turned out badly due to, say, another driver pulling out unexpectedly without looking. In this case, she claims to have acted carefully, but that due to events beyond her control and about which she had *no knowledge* at the time, an accident occurred. If this were to be shown to be the case in the subsequent investigation of the timeline leading up to the crash, then she may well be held to be blameless.

---

[1] On these different senses of responsibility, see for example Ingarden (1970: 5–34), and Birnbacher's (2001) distinction between retrospective responsibility, or '*ex post* responsibility', and responsibility for the future, or '*ex ante* responsibility'.

Here we see that liability and blame depend on knowledge: first, knowledge of a timeline of events leading up to damage being done, and secondly, knowledge on the part of whoever has played a causal role in this timeline. If the driver can be assumed to have known that she should not have pulled out from the junction when she did, then she would be to blame and legally liable. If this assumption does not hold, then she would be morally and legally innocent. However, if we shift the context to that of the knowledge practices we have analysed in previous chapters, then the first three ways of understanding responsibility we mentioned (holding someone responsible, assuming responsibility, being responsible) as blame and/or liability appear to be hindrances to addressing harm rather than useful tools.

Reliance on industrial use of nuclear, bio-, geno- and nanotechnology, together with economics and liberal democratic politics, all encourage us to fly blindly forward into the future, trusting in the protection of foresight* and scientific prediction. As noted earlier in Futures Traversed, the practices through which these institutions construct futures effectively institutionalise irresponsibility, exploiting the future in the narrow interests of the present. Whether it is instability arising from decisions taken by parliaments elected for five-year terms, economic chaos arising from the aggregate effects of millions of individual quests for quick profits, or the unintended consequences of industrial-scale penetration into the basic structure of organic and inorganic matter, short-termism in human action tends to produce effects that could never have been foreseen in the laboratory, boardroom, or cabinet office. The irresponsibility here appears to be structural*, because uncertainty is built into the practices that inform and shape human action. It is this social and institutional context that has gradually come adrift from the moral and legal practices that have historically accompanied and moulded human action (Adam 1998: ch. 5; Pellizzoni 2004: 553).

The problem concerns the practices by which we *know* the future, and the assumptions on which they rest. Scientific and economic practices, as previous chapters have detailed, work on the basis of deterministic natural laws, and construct abstract futures* on this basis by relying on knowledge of the past for predicting the future. However, the complex realities of technological, economic and political practices mean that they feed into natural and social processes in ways which create thoroughly unpredictable and irreversible effects. Further, as they flow into these wider systems, cutting across different levels of reality, they alter the course of emerging, living futures on the way. Consequently, the more complex the interactions that knowledge practices unleash, the more likely it is that their effects will not emerge for long periods of time—as is clearly shown in the case of the bioaccumulation of synthetic chemicals (Colborn et al. 1996; Koppe and Keys 2001). When latent*, living futures that imply complex, multi-level interactions between different kinds of phenomena are created in this manner, practices of moral and legal judgement run into two sorts of difficulties.

First, they are tasked with assigning responsibility where harm occurs. But where damage is done, they find themselves having to deal with situations that are structurally similar to that of the driver who claims that, based on what she could have been aware of at the time of the crash, she could not have taken action to avoid causing what subsequently occurred. Because of the way responsibility is typically understood as being rooted in causal *authorship* of harm, it is easy to avoid responsibility by pointing out that the risk of causing a particular harm (such as the loss of fertility in animals and humans caused by bioaccumulation of PCBs, or the financial chaos following the collapse of a hedge fund) could not have been predicted at the time of acting, given the then-current state of knowledge in science, economics etc (Pellizzoni 2004: 552). This recourse is open to technologists and politicians alike. Nonetheless, because of the open-ended timeprint of the knowledge practices that exploit empty and predict abstract futures, there follow inherently unpredictable consequences. As previously noted, this means that irresponsibility is inherent in the construction and production of long-term futures. It is this *structural* irresponsibility with which traditional legal and moral understandings of responsibility have trouble.

Secondly, because of the focus of legal and moral judgement on assigning blame for harms that have *in fact* happened, no account is taken of the need for understanding what might constitute action that takes responsibility *before* the fact. Law provides norms of public accountability, and as such is concerned with the public recognition of wrongs that have been done. But in the action sphere with which we are concerned, it is the connection between actions that are, at present, legitimate in the eyes of the law and future harms that is the problem. When the ever-faster production of empty futures is facilitated by technologies of travel, transmission and transplantation, then the boundary between present acts and future harms becomes ever harder to draw. Again, the *structurally irresponsible* tendency to produce futures with long-term latency that implicitly take futures from others is the problem. The difficulty with the traditional legal and moral model of responsibility is that the ethical significance of the ways we now produce futures are invisible to it.

Of course, moral practices have sometimes looked forwards, by establishing general principles of conduct for action now and in the future, or by describing what counts and will count as good practical judgement. Similarly, some areas of legal practice, such as constitutional or trust law, have concerned themselves with the future security of legal contracts. Our social memory includes these and other modes of future-orientation, without their inner core being made fully explicit. This core is the shift in attitude that occurs when we view responsibility as something we actively *take* rather than something which is imputed to us when we are held liable for our actions. In this chapter, we will explore how this forward-looking orientation can enable us to understand responsibility as being, first and foremost, about a responsible 'handling' of the world (Ingarden 1970: 14–17). Before we do this, however, we need to outline some of the conceptual supports for the legal and moral habits of mind* we have traced here, in order to better understand the deeply ingrained nature of what must be overcome.

Limits to Reciprocity

These conceptual supports concern the basis of law itself in democratic polities, namely the nature of the social contract which is enshrined in civil and criminal law. They reflect, we shall suggest, assumptions about the nature of moral agency and the obligations that accompany it that lend themselves to the construction of empty futures. This inherent tendency is a result of the temporal bias towards the protection of the present which is built into these assumptions, and which we shall now examine. As we shall see, it meshes with the practices of economics that we analysed in Futures Traded.

Hans Jonas (1984/1976) has argued that for much of human history the meaning and limits of moral and legal discourse reflected the spatial and temporal boundaries of human action. Where the effects of our deeds are limited to immediate contemporaries with whom we share public space, certain specific concepts of duty, blame and liability are required. Jonas (1984/1976: 3–7) writes that from ancient Greece to the European societies of the 17th and 18th centuries, ethical theory held that the mostly face-to-face nature of interactions between people meant it was important to establish who was to blame for specific harmful actions.

This ethical context was also reflected by the liberal republican tradition in political philosophy from John Locke (1632–1704) to John Rawls (1921–2002) and beyond, which sees the basis of social order as a social contract that exists between inhabitants of a locality, or at most, a nation or people with a particular constitution and body of laws. However, when the context of action extends to include people on the other side of the world, near and distant future generations, and the natural systems needed to sustain them, then, Jonas suggests, such concepts are in need of revision. To interpret the social contract as incorporating solely those who belong within a single national community is no longer appropriate.

The reasons why the traditional conception of a social contract is no longer suitable in this new ethical context have to do with the way its inherent temporal bias results in the inadequate legal and moral concepts of responsibility we have already examined. In addition to the problem that legal accountability, as we have seen, is based on knowledge of connections between past and present, the social contract has at its root the concepts of *reciprocity* and *autonomy*, both of which tend to restrict our ethical vision to the present.

Let us see why this is the case. The basis of the rule of law is generally taken to be its equal applicability to all. This is rooted in a fundamental innovation of the liberal democratic tradition, that is, the formal and natural equality of all human beings.

> And, being furnished with like faculties, sharing all in one community of Nature, there cannot be supposed any such subordination among us that may authorise us to destroy one another, as if we were made for one another's uses, as the inferior ranks of creatures are for ours. (Locke 1988/1689: 2nd Treatise, ch. 2, §6)

If there is no legitimate reason to insist on a natural hierarchy in nature (e.g. of free citizens over slaves, Europeans over non-Europeans, or men over women) then a basic moral responsibility to respect the natural equality of all exists, independently of the existence of actual laws. Formal equality before the law rests on the equal moral entitlement of all individuals to have their humanity respected. This entitlement generates a reciprocal relationship between all individuals who are capable of harming each other. Each is enjoined by a basic moral 'law of nature' to respect the intrinsic value of every other person, which brings us to the role of the concept of *autonomy*, from which this intrinsic value is held to derive. Moral autonomy is thought of in this tradition as the ground of both the natural equality and the dignity of human beings. As described by the psychologist Lawrence Kohlberg (1981), autonomy is a cognitive achievement, the ability to distinguish right from wrong by disinterestedly and impartially applying objective moral rules to situations. It involves the capacity to decide for ourselves what rules of conduct we should follow, and implies that we are free to have done otherwise than how we actually acted. It is therefore the rational, cognitive side of human nature that the liberal democratic tradition holds to be the source of the equal intrinsic value of individuals.[2]

When we look at these ideas about the basis of moral and legal equality in the light of the changed context of action that we have outlined in previous chapters, then several problems become apparent. In a context in which human actions create latent, living futures that draw into them both the futures of human social systems and those of natural systems, the question of how these futures *matter* to us is of crucial importance, as it is only if they do actually matter to us that we can take responsibility for them. As we saw in the last chapter, the future is the temporal dimension of experience through which meaning is projected and woven with the past and present. But traditional social contract theory makes its focus the present and near future of individual lives. For this theory, my responsibilities to other individuals are in the main to other living moral agents whom I could harm directly and who could harm me. The futures that matter from this moral perspective are ones encompassed within the basic framework of the social contract: futures falling within the horizon of an individual lifetime that affect the rights of living individuals. Consequently this perspective is synchronised with that of the other present-focused practices we have analysed in previous chapters.

Indeed, other knowledge practices that link social contract-based moral discourse with economics and public policy demonstrate how mutually supportive they are. We saw in Futures Traded how economics constructs the future as an empty, quantifiable medium and uses it as a tool for assessing the costs and benefits of different actions in the present. To

[2] The connection between autonomy and rule-following was decisively made by Immanuel Kant (1993/1785: 40–1). Roderick Chisholm (1967) defends the view that autonomy implies the choice to act, and Christine Korsgaard (1996) articulates the connection between autonomy and dignity.

base law on the social contract, and thereby upon the equal responsibility of all to respect each other, buttresses this construction of empty futures. This is because the principle of equal respect marks out a zone of protection for each individual and their interests. It establishes that all have a right to pursue their interests without being harmed, so long as they do not harm any other living person. Economics provides a host of social practices through which individuals can plan how to satisfy their interests by saving, investing and so on. The idea of equal respect establishes a moral and legal basis for the construction of empty futures by individuals and the groups of which they are members, including private companies, equity funds, unions but also groups with wider membership, such as all those who pay tax within a nation.

Amongst the social practices provided by economics and supported by the idea of reciprocal and equal respect are cost benefit analysis (CBA) and future-discounting*, as mentioned in Futures Traded. Outlining how these tools work reveals the extent to which the conjunction of empty futures and the principle of equality for living individuals collude in pushing the future beyond consideration. Using CBA means assigning to an action, policy or product a set of financial estimates of the costs and benefits it will produce, in order to determine whether it is of net positive or negative value. To allocate values, practices of 'future discounting' are used to reflect the assumed preference of living people for obtaining what they desire more quickly. For example, the value of £100 held now is seen as greater than the same £100 in, say, ten years. This is true whether the £100 is a cost or a benefit. Consequently, it is assumed that an individual would much rather put off *paying* £100, but, if receiving the same sum, would prefer to get it sooner rather than later. Similarly, if we have the right to use a particular resource, then to claim our rights now would realise more value for us than putting off exercising them (Jacobs 1991: 68–70, 81–82).

Given that the right to pursue one's interests is assumed to be the equal right of all living individuals, then moral and legal practices join these economic ones in protecting the interests of the present directly at the expense of the future. In relation to these practices, it makes complete sense to, for example, employ lightly regulated complex technologies in the present to achieve results as fast as possible. It is therefore much easier for courses of action to be approved that have a high probability of short term benefits and any probability, whether low, high or uncertain, of long-term costs (including the ill-health of others, disease and resource deterioration). The meaning of action, the way it weaves together past and present with the future it tries to create, is fixed in relation to short-term goals only, and to the assumption of the equal right of living individuals to pursue their present interests.

Having reviewed how economics and moral reasoning together make only the empty, short-term future an object of concern for us, we shall now briefly examine how this framework erodes any concern for other futures. The principle of equality between autonomous human beings excludes, as numerous philosophers have noted, both non-humans (present and future) and future humans from the 'community of rights'. To possess moral and legal rights, we must be capable of either exercising them or of appointing someone to do so on our behalf (Jonas 1984/1976: 38–39; MacLean 1983: 183–4; Steiner 1983: 154). Animals, plants, ecosystems and potential humans can do neither. But the context in which the principle of equality and its legal expressions hold true has changed. Structural irresponsibility now extends the reach of harm out beyond the 'community of rights' to encompass all these excluded categories. Without reflecting this shift, the whole sphere of morality and law must remain tied to the present and its interests.

To sum up, we have seen that the principle of equality between living, morally autonomous persons establishes the moral basis for their legal equality and is expressed in legal concepts of harm. To harm another is to transgress against the rights they possess exclusively by virtue of their humanity. As noted above, this definition of harm excludes the whole of nature from moral consideration. Furthermore, a key feature of reciprocal responsibility is a bias towards the present. If we link this bias within the moral and legal concepts that establish guiding norms for action to that of economic practices like CBA, we see that the fundamental intellectual currents that support and ratify legal and political institutions also support and ratify the exploitation of the future. Together with the past-based focus of the concept of accountability, which requires that harm has been caused and that its causal authorship has been established, these basic assumptions form a formidable barrier to any appreciation of why futures matter.

Care Comes First

In Futures Traversed, we noted that to rebuild the links between action, knowledge and ethics requires that we rethink ethics in its relationship to action and knowledge, especially in contexts of uncertainty. In Futures

Thought, we suggested that the future matters to us because of the link between the future as *living* and as *lived.* Through futurity, unforeseeable novelty is introduced into our world. On the basis of our experience of it, meaning is produced and the stories of our lives woven.

According to the idea of a social contract, all autonomous individuals have reciprocal duties to each other because they could choose to act in a way that ignores other peoples' rights. But in relation to the future this leaves out an important consideration: we are all already involved with the future as its creators. When we extend ourselves into the future through imagination and through action we make and take futures. Because this is the case, there is a basic inequality of power between present and future that does not exist between living contemporaries. In a globalised world, it is beginning to be recognised that both the ecological footprint of our lives and what might be termed the social footprint need to be taken into account. Through our actions, we are inextricably connected with both ecological support systems and people across the world. The benefits we gain from economic and social institutions are often extracted from these relationships to the cost of both natural systems and distant people we will never meet, as Iris Marion Young (2006) argues. But this interconnectedness also holds, in terms of our temporal reach, the timeprint of our actions. If we act solely with an eye to the meaning of our actions for our present interests, then we tend not to be able to comprehend anything of the ways in which they are already generating latent futures which we will not live to experience. To encompass the extent of the timeprint of our actions we need to understand more of how we are involved with the future, and how this involvement *matters* to us in the ethical sense, i.e. as a source of guidance about what we *should* do. This, as we shall see, inevitably leads us towards what Jonas (1984/1976: 94) calls a non-reciprocal* concept of responsibility for the future.

This involvement is, as we saw in the last chapter, constitutive of what we are, at the chemical, biological, psychological, and social levels of our existence: it is both our lived and living futures. But the principle of reciprocal responsibility and the standard ideas of liability and accountability work to separate the present—as if it were a distinct spatial zone, an expanded present moment—from its involvement with the future. We need an understanding of the new ethical context described by Jonas that overcomes this separation. As detailed in Futures Thought, our starting point is to return to the description of the lived future given by Heidegger, in which he calls our involvement with the world *care**. It is this lived future that connects us with living futures whose potential for emergence extends far beyond the span of our own lives.

When Heidegger says that humans 'are' care, what he means is that we do not encounter the world in the same way as, say, a rock or a horse. A rock is subject to physical forces that might make it roll down a hillside and strike another rock. A horse is driven by a kind of biological anticipative concern for its own interests which, at some level, we also share—this is the lived future of Jonas' *conatus**. Care on the other hand is the way in which the world takes on significance for us in relation to our *interpretation* of our interests and the future horizons they foreshadow for us, an activity which unites our emotional, imaginative and rational sides. This motivates us in ways that are not simply reducible to survival instincts, because it is related, first and foremost, to what we feel the ethical meaning of our lives to be. The continual reaching beyond what we *are* to explore what we *might* become is the motor that generates the narrative structure of our lives and gives them a kind of unity over time. Things and people matter to us and take on ethical significance because of their place and role in this story, both of which may also change with time. It is through our imaginative and emotional concern for what their potential might hold for us in the future that what they mean for us now becomes apparent.

It is important to remember that care in this sense is inescapably a social process, one conducted through our relationships with other people. Our involvement with the future is therefore always involvement *with* others. Some developmental psychologists argue that from the earliest stages of our development onward we are to some degree conscious of being with others who are like us but different, as this dyadic relation is a condition of our own evolving sense of self (Benjamin 1988; Stern 1985). In this sense, we do not only care about the meaning of our own lives; we care also about the significant other people (parents, siblings, friends, lovers, colleagues etc.) with whom we share our everyday world. Our relationships with significant others, as we grow up, are the medium through which we learn about the world and ourselves. It is by having such relationships that we first learn about ethics, about the fact that there are things that we *should* do for their own sake. This form of responsibility is one that specific situations call forth. Moreover, as we get older, experience hones our judgements about what we need to do in different contexts. This form of responsibility requires that we *tend* to a relationship by providing what is needed by another person or persons. Instead of requiring us to *refrain* from doing something, as in the case of reciprocal responsibility, it actually encourages us to extend our zone of action in order to conserve and sustain relationships (Gilligan 1982: 38). To care in this sense therefore means to take on a non-reciprocal responsibility for performing a particular task because it falls uniquely to us to perform it.

This is quite different from the equal and reciprocal responsibility to respect the intrinsic value of other people. First, it is not necessarily equal. A parent has specific responsibilities to a child, but the child does not share these. Such responsibilities are a product of the distinct relationship that exists between them, and are undertaken in order to sustain the relationship itself. Similarly, someone would give support and sympathy to a friend because of their friendship, and in order to help their relationship flourish. In either case, the value of the relationship is the key factor in motivating responsible action, and is also the object of acting responsibly. The social meaning of being a parent implies that the welfare of the child is of primary importance, and that this is central to any parent-child relationship. Similarly, being a friend implies that the welfare of our friend is of primary importance. In either case, we act not because of a sense that the other person is of equal value to ourselves, but because they are of *special* and *unique* value to us. This also means that the other is not just anyone, which contrasts with the presumption of equality, where the specific identity of people is of no importance to their legal status. Care acknowledges that moral significance attaches to someone's specific relationship to us.

Secondly, all relationships of care are conducted *in time*. That is to say, they are not static relationships that connect us, at all times and in the same way, with others. They are always extending into the future, towards a horizon of what Harry Frankfurt (1982) refers to as *common fate*. When we care about another, we acknowledge that their future, their welfare and ethical significance are bound up inextricably with our own.

Care is therefore both future-directed and always attached to specific individuals. It is thereby specifically directed towards their futures, and is therefore tied to futures which are embedded in distinct contexts of concern. Consequently, it constructs lived and living futures. In caring for another, we attempt to judge what futures they project for themselves—what they want and need—and what they are becoming, both because of what they want and need and in spite of it. What we attend to is the unfolding potential of an individual, and to what events will mean in the context of the fate we share with those we care about. The orientation of caring is therefore very different to that associated with moral autonomy in Kohlberg's work (1981). Kohlberg's definition of autonomy implies the capacity to transcend the specific attachments implied by caring relationships and judge impartially, applying a single moral principle to different situations. This is represented as a purely rational cognitive achievement. Marilyn Friedman (1993: ch. 1) notes that, given the nature of human attachments, it is difficult to see how such a form of autonomy is in practice possible. Further, to care effectively for others involves all our capacities, producing finely discriminating judgements in sometimes entirely novel circumstances, all of which are interwoven with our need for meaning (Nussbaum 1990: 68–72). Care is a product of experiences, including the loss of intrinsically irreplaceable relationships, which can only result from attachments to specific individuals.

We now will end this section by taking further this comparison of the two perspectives. As we saw, the autonomy perspective sees the boundaries between living individuals as crucial for understanding which actions are of moral and legal significance. The consequences of this include a tendency for the future to become morally invisible for actual social practices. We can now see why this is so: the autonomy perspective attaches value to the dignity of living individuals because they represent singular *instances* of something which human beings are held to generally possess, namely moral autonomy. But the problem is, as we also saw, that this universal source of respect and value is only present in specific living individuals. This means that assumptions about autonomy tend in practice to *isolate* the present from the future. Caring, by contrast, acknowledges the way in which the present is always involved with the future, in the weaving of a common fate with others.

Autonomy and reciprocity are, in the liberal democratic tradition, seen as the logical foundations of the social contract, the moral basis for the rule of law. This contract, however, holds between living individuals, each of whom should choose to respect the rights of others. Care, by contrast, is the existential *substance* of the social contract. It is through our empathy and identification with concrete others that harm matters to us morally in the first place. Further, as we shall now see, this substance of our common ethical and political life embraces futures beyond the lifespan of living individuals who feel they share a 'common fate'. It may appear that we can only care for those of our contemporaries with whom we share a special emotional bond. It is thought that such thick bonds cannot be formed between living humans and either future humans whom they will never meet or non-humans. However, although our capacity for care may not be able to embrace all others in the same direct way it does those who are close to us, understanding human beings as 'caring' subjects gives us a different perspective on the meaning of ethics. This is directly linked, as we have seen in this section, to the special ethical significance of futures. We will now develop this perspective further by looking at how care can be thought of as extending beyond the circle of our closest relationships.

Common Fates

In suggesting that the existential role of care is the basis of ethical engagement with others, we have effectively reversed the priorities of the liberal democratic tradition. Connectedness and the specific weights accorded to concrete relationships appear in this context to be more important in moral judgement than the abstract rightness of principles. The value of principles would, from this perspective, be dependent on how far they help people live ethically meaningful lives. This changes the whole model of ethics which we employ in order to re-connect with the future, and will need to alter its legal expressions too, as we shall see in the next chapter.

An example of this difference between two ethical perspectives is provided by Tzvetan Todorov's (1996) account of how men and women responded to being incarcerated in Nazi and Soviet concentration camps. Whereas men tended to separate from each other, striving to maintain a sense of dignity in the face of their sudden vulnerability to brutality and humiliation, women found countless everyday practical ways of caring for each other (Todorov 1996: 77–8). Todorov (1996: 77) observes that "on the whole, women survived the camps better than men did, not just in terms of numbers, but in terms of psychological well-being".

If we relate this example to our previous discussion, we can see the importance of this difference in perspectives. It suggests that our sense of the ethical significance of our lives, together with our wellbeing, tends to be connected to whether or not our capacity for care is allowed to express itself. In the terminology of moral philosophy, this connectedness with others is *constitutively* valuable*. That is, the concrete relationships we have with others are essential ingredients of our wellbeing. Two features of this way of being valuable are important.

First, such relationships have a specific kind of structure. Our friends are constitutively valuable to us because we feel that without them, our lives would lack something vital for our wellbeing. But friends can only be valuable in this way so long as they exist and flourish in their own right as individuals. If we exploit a friend for our own ends, they cease to be valuable in this way. They become only *instrumentally* valuable. The meaning of our relationship changes, and may be irretrievably damaged as a result. Consequently, their life and wellbeing have a specific weight of meaning for us, for without them doing well and being happy in their own right we would feel diminished. Secondly, just as care in general is concerned, above all, with the lived and living futures of those we care about, their constitutive value also derives from their futurity. The fact that we feel them to be involved with us in a 'common fate' gives them this specific kind of value. They matter to us because of their potential, which implicates them alongside us in living futures that have yet to emerge.

As the constitutive value of someone derives from their ongoing contribution to the meaningfulness of our lives, care for them connects us directly with our lived experience of the future. It also however necessarily connects this lived experience with living futures that exceed our own lifespans, the latent futures of processes* that may take hundreds or thousands of individual lifespans to coalesce into products. This is because our care is inevitably also an evolving and extending network of relationships, and one that, due to its inherent future orientation, opens onto future horizons of potential that stretch out much further than those which frame the narratives of our individual lives. We now look in more detail at how this network extends beyond the boundaries of our connections with individuals in the present.

First, we do not only care about other human beings who share our past and future with us. Our concern for ethical meaning opens itself out in many different directions, extending to connect with many kinds of objects that we also find to be of constitutive value. An artwork, a landscape, an institution, an idea—these are a few of the things we can value and care about in this way. Just as it is important to us that our friends, lovers and relatives flourish and enjoy wellbeing, so it is important to us that something we find beautiful, profound, enlivening or inspiring should be helped to flourish as well. When we care about people, we care about and strive to accompany them into their individual and collective futures. We seek to involve ourselves with other objects of constitutive value in the same way. In doing so, our aim is not simply to preserve something in stasis, just as we do not hope that a friend will somehow simply remain the same as they are now. In caring for things as diverse as natural habitats, democratic institutions, or the works of Beethoven, we are concerned that they should continue to be sources of meaningfulness for ourselves, others with whom we share the world now, and future others whom we shall never meet. Harm done to such objects can be experienced in a similar way to harm done to significant others: it also diminishes us.

Secondly, our existential care perspective necessarily builds for us a complex interlinked moral world in which (for example) to care about a loved one means also being committed to how they are looked after when sick. This might lead us to campaign for the public provision of healthcare and to join a political party which makes this an essential part of its campaigning platform. Caring cannot help but involve us in cultivating new relationships that we then also find to be constitutive of both *our* wellbeing and that of all those (human and non-human) about whom we care. This process may lead us to revisit and revise our previous commitments. In extending care, we have to assess, based on experience, the evolving systemic relationships between plural values over time. Consequently, our private caring necessarily reaches out into the public world, and to the natural world beyond that.

From the point of view we are describing, moral development does not lead via a linear series of steps towards Kohlberg's (1981) ideal of autonomy, but instead gradually evolves into a complex multi-level and systemic view of how our constitutive values fit together. From our caring relationships with individual people, we extend ourselves to an intersubjective context, and from here, to a social, institutional and historical one. From this, as our contemporary situation makes clear, we must move to an environmental and ecological context, in which landscapes, biodiversity and other aspects of our ecology are recognised as also being constitutive of our wellbeing now and in the future. To allow, and perhaps to help, natural phenomena to flourish forms a necessary part of the circle of our caring as much as attending to a friend does.[3]

Each level of interrelation changes our understanding of the ethical meaningfulness of our lives, by seeing these lives within the context

[3] For an alternative view on how we can enshrine the value of nature in moral and legal practices, see Stone (1996).

of new future horizons that stretch out beyond our own lifespans. *The individual lived future that is at the centre of Heidegger's work, we want to suggest, should expand our ethical perspective and connects us with both the lived futures of other individuals, and the living futures in which these other perspectives participate and to which they contribute.*

The meaningfulness of our own lives as such is bound up with things, people, institutions and ideals. This means that our lives can be said to continue to go well or badly even beyond our own deaths, because of what might happen to these objects of value. So, for example, if we became convinced that our grandchildren would be forced to live on an earth devastated by war or climate change, then our own lives now would fall under a shadow. Similarly, if we are scientists, and were to become convinced that the distant future would see the institutions of science, and with them, scientific knowledge itself largely vanish, then we would also feel diminished by the prospect. Part of our sense of whether or not our own lives are going well now is bound up with our expectations for the ongoing futures of what we care about, even beyond our own deaths. In this sense, as John O'Neill argues (1993: ch. 3), we feel that how the future turns out can add to or subtract from the value and meaning of our own lives. The stronger this sense of participating in projects which connect us with future generations, the stronger is our sense that near and distant futures both matter to us now. Our common fate necessarily weaves our individual stories into those of our descendants near and distant, just as the actions of our ancestors wove the longer threads of our and their common fate, to which we have added our contributions. Constructing our own futures through imagination and action forges novel connections that in turn unleash living futures that far outlive us. The perspective of care sensitizes us to how these futures interweave, and alerts us to the need to handle responsibly a world that is spun from myriad relationships and commitments, attending all the time to what the things we care about need from us in order to continue to flourish.

Although care is rooted in our singular lived experience of the future, it necessarily leads us towards some general principles that should inform action in the present in order to safeguard the future. First, given the nature of our technologically enhanced power to create futures, the primary general rule we should use to guide our individual and institutional activities is, as Jonas has written (1984/1976: 37), "never must the existence or the essence of man as a whole be made a stake in the hazards of action". If we adopt a perspective that sees care as the basis of the relationships that constitute social life, this is a necessary conclusion, and one derived not from fear of the consequences of irresponsibility, but from a positive desire to tend to the wellbeing of what matters to us. Here we differ from Jonas, who (1984/1976: 26–7) sees fear of the destruction of humanity (whether imminent or more distant) arising from futurologists' hypothetical scenarios about the future as the primary motivation for adopting a new future-oriented ethic of responsibility.[4]

Secondly, our ramified strands of care must necessarily connect with each of the personal, social, ecological and historical dimensions we have outlined above. At all of these levels, care is exercised for the embedded futures of what and whom we care about, reaching finally towards posterity, and the living futures that are already underway within our actions in the present. If posterity matters to us in this way, then what really concerns us, first of all, is that the human race should continue to exist with the capacity to care and to find life meaningful. The futures about which we care are, in this sense, the futures of care itself.

Thirdly, this ultimate concern means we have to specify a *positive vision* of what a good quality of life consists in. Manfred Max-Neef (1992) has articulated such a vision in terms of an interlocking and evolving system of human needs, which includes within it both means of subsistence and sources of meaningfulness. To make this move means that we have to go beyond the liberal democratic vision of morality and legality (as being first and foremost about the defence of individual autonomy) and defend an ethical vision that places 'the good life' at its heart.[5]

If what we think of as good depends on what we care about, we might seem to face a problem here, because others might not care for the same things we do. However, this is not a serious objection. It is true that our care for the future is initially experienced through our specific connections to things and people here in the present. Nonetheless, once this care perspective opens up, as we have described, and extends itself further into the future (as, we have argued, it necessarily must), it has to encompass more general concerns about what is of constitutive value to human beings. It is difficult to imagine that there will be absolutely

[4] As Richard Wolin (2001: ch. 5) has noted, this adoption of the "heuristics of fear" influences Jonas' arguments later on in his book for authoritarian and anti-democratic institutional solutions to the problem of irresponsibility.

[5] See Henderson (1988) for an example of how a similar view of value can be used to develop a new basis for economics.

no room for agreement between individuals, groups and nations here. As Max-Neef (1992: 199–200) argues, what will count for different people in different times and places as satisfying diverse human needs cannot be predicted. But there is perhaps no fundamental obstacle to trying to describe the needs themselves.

Further, the potential for change in what is cared about is built into care itself, and the preservation of this potential must also be of central concern. As Jonas has pointed out, at the centre of any vision of the good life must be the need to preserve the capacity for care itself, including the capacity to change what is cared about. The scientist in our earlier example would not want to pass on a scientific tradition that was dogmatic and immobile—one without, in effect, any kind of living future and potential for evolving meaning through dissent, criticism and experimentation.

In recent times, responsibility for the future has been articulated chiefly in terms of the precautionary principle, i.e. that action should be prevented where

> the level of scientific uncertainty about the consequences or likelihoods of the risk is such that the best available scientific advice cannot assess the risk with sufficient confidence to inform decision-making. (UK Government Interdepartmental Liaison Group on Risk Assessment 2002: 6)

However, taking a care perspective means that precautionary action would be made appropriate wherever the degree of scientific uncertainty signifies that the risk to future quality of life cannot be assessed. In other words, the *negative* measure of refraining from action would be motivated by a positive vision of the good life. This would hopefully mean that the application of the precautionary principle would extend beyond the areas in which it is most used now, such as environmental and food regulation, to encompass, for example, the impact of technologies on social equality and health. It would also hopefully mean that its application would be easier, given that a list of specific, concrete needs and means of realising them would be provided in relation to which risks could be assessed.

Reflections

We have argued that, by changing the way we think about the basis of ethics from autonomy and reciprocity to care and non-reciprocity, the future can be brought to the forefront of our concern. In the contemporary situation, the boundaries between present actions and future harms are hard to draw, given our collective capacity for creating long-term latent futures. By focusing on the constitutive value of what matters to us, and therefore on its embedded future(s), concern for care seeks to protect and enhance potential, and to create the social infrastructure that will enable people to accompany their legacies into the future. The perspective of care reflects an artisan's point of view on futures, where responsibility is always an integrated element of the relationships through which the form and meaning of the world are produced. The potential of these relationships, their living futures, far exceeds what we can now see in them or draw out of them. Ultimately, it is this potential that the imperative we introduced in the previous section, which demands that we *care for the future of caring itself*, seeks to safeguard and sustain.

CHAPTER NINE

FUTURES TRANSCENDED

Introduction

With this book we have sought to lay foundations for approaches to the future that are appropriate to their contemporary context of future making. While we fully appreciate that there is much work still to be done, we are also aware that this exploration has enabled us to make the future tangible, render the invisible visible. This in turn affected the way we understand our role and our implication in potential effects. How we see our actions, knowledge and responsibility has irreversibly changed. Moreover, there is no going back. Just as we cannot re-enter the pre-linguistic world of the infant after we have acquired speech, so we cannot return to a two-dimensional futures world of space and matter after we have begun to integrate temporality and futurity into our knowledge practices*. As readers you have joined us in this endeavour, which means that you too have passed a point of no return. Your world too has irrevocably changed and these changes reach deep. Thus, for example, once temporality and futurity are explicitly encompassed in our knowledge practices, their negation becomes a consciously willed act. Similarly, any denial of implication in potential outcomes and the exclusion of latency periods* from horizons of concern will require active effort. The designation of either as unreal will be based on choice rather than ignorance. What used to be implicit and taken-for-granted has become illuminated, explicated and transformed into subjects of reflection. And with this shift in standpoint and perspective the foundations are laid for new knowledge practices: each one of us is charged to help create on these foundations structures that are appropriate to both their base and their contemporary context.

Since the laying of these foundations has been a joint effort involving authors and readers, the resulting structures will differ greatly, as each must suit not only its unique context but also its inhabitants, who have their own diverse biographies and needs. Every participant in the exploration will have drawn different inferences from the processes we investigated and the stories we told, will have identified different priorities and mapped out different solutions. As authors we set out to find, mark and map access points for change. Making changes, however, requires context-specific practices that will inescapably differ not only with the level at which action is conducted, the competence and the skills applied but also with the associated sphere of influence. None-theless, despite our awareness as authors that every reader will need to appropriate the insights gained for *their* own purposes, construct the new buildings according to *their* requirements and make changes appropriate to *their* context, there is still one knowledge domain that requires our further attention. There is a need to re-visit the conceptual tools we have touched upon throughout this text and focus once more on the relationships between them. The tools in question are the concepts and frameworks of meaning that constitute the scaffolds for this undertak-ing. Not quite enough has been said about the relations between the components to consider this part of the exploration adequately covered. Some of the conceptual issues already addressed we would like to develop further, others we would like to weave together to give them additional strength and solidity. Appropriate conceptual tools, we want to argue, are essential for re-building approaches to the future. The further development of such tools is therefore the endeavour with which we want to bring this book to a close. 'Tools' sounds frightfully utilitarian and technical, yet the objective is not technical upskilling. Rather, the purpose is one of relating and integrating, of binding into a coherent whole the fragmented domains of action and being-becoming that have been sketched in previous chapters.

Taken-for granted assumptions are invisible. They belong to the world of know-how rather than explicit knowledge. If, as 'common sense', they act invisibly as barriers to desired change then we first need to render them visible. Once habits of mind* are brought to the forefront of consciousness and we learn to 'see' them, we can begin the difficult task of changing those deeply sedimented ways of knowing. As with all the other issues we discussed in this book there are two key insights. The first is that *things could and can be different.* The second is that *our extensive thought traditions are not lost but enfolded in who we are and what we know.* It is up to us therefore to recover these traditions and consider their (in)appropriateness for contemporary *future matters.* We should ask whether and in what way they might be usefully adapted for our contemporary context, and what difference it might make if parts of this enfolded ancient knowledge were unfolded, recovered and adapted for responsible future making. Again, this is not about going back. Rather, it is about moving forward with an expanded consciousness, emboldened by the knowledge that the industrial way of life is not destiny: there have been and still are other ways of living our lives.

In the first section of this final chapter we reconnect what has come adrift during the industrial way of life: the production of futures, knowledge of futures and our responsibility for potential outcomes. This means relating action, knowledge and ethics. Since everyone has different competencies in these domains of the future, we explore a number of access points where improvements could be achieved and consider what these might entail. Next we revisit key assumptions that tend to predispose current knowledge practices towards a lack of concern for long-term effects of contemporary future making. We focus on deeply sedimented habits of mind and, in some cases, trace their cultural roots. Our aim is to identify openings for change, ones which may produce approaches to the future that are adequate to our contemporary timeprint*. What we want to transcend, therefore, are not futures but contemporary perspectives and approaches to futures and futurity.

Action, Knowledge, Ethics

In the course of this investigation one thing has becomes obvious: making futures is easy. Everyone does it all the time and with great facility. Knowing these futures with all their impacts and ramifications as they stretch across time and space, in contrast, is impossible for all but the most repetitive of actions and events. This disjunction between action and knowledge has implications for the way responsibility for future effects is approached. As long as responsibility is tied exclusively to known outcomes of policies, actions and inactions and excludes impacts that are shrouded in uncertainty, futures will continue to be produced with impunity. See Figure 2.

The three interconnected elements of action, knowledge and responsibility do not play equal roles in our contemporary relations to the future. Often they are treated quite separately, having come adrift in our world of compartmentalised knowledge. If we can agree that those three spheres belong together and that future making ought to be done knowledgeably *and* with responsibility then we need to understand the reasons for the disconnections as these may help us to find ways of reconnecting what has become separated with the industrial way of life and re-align the three elements in accordance with specific contexts.

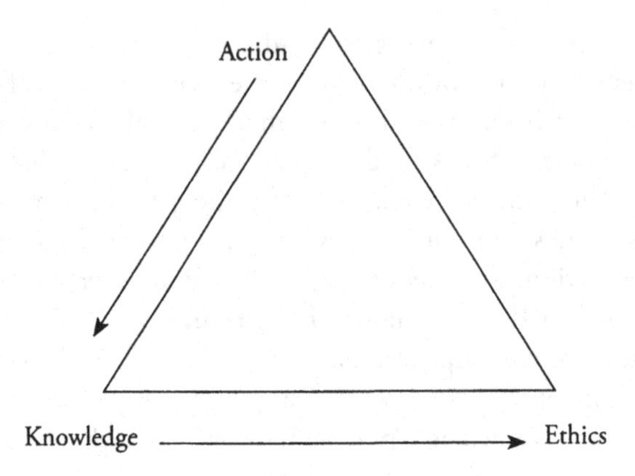

Figure 2: Ethics Based on Knowledge

In the first two chapters of this book we suggested that it makes a difference to our action potential whether the future is conceived as pre-given and actual, as empty* possibility, or as process* realm of latent futures in the making*. Who owns the future, we argued further, has knock-on effects for the way it is perceived, the nature of the knowing and the anchoring of responsibility. Thus, efforts to *know* the future are more likely to involve discovery, disclosure and interpretation of destiny, fate and fortune if the future belongs to god(s). If it is tied to the cosmos, in contrast, then calculation, prediction and extrapolation of planetary movements and auspicious moments for change may be involved. But, if the future is seen as ours for the making and taking, then imagination may be employed for conjecture, creation, colonization and control. With respect to the interdependence of action, knowledge and ethics we argued that once people's approach to the future shifts from seeing themselves as recipients to understanding themselves as protagonists and agents of change, the locus of responsibility changes too. The onus is then on the new future makers to know their productions together with their potential ramifications. This, however, is far easier to demand than it is to achieve since the changing locus of control is accompanied by massively increased uncertainty. That is to say, when the future is no longer thought to pre-exist but is approached instead as a realm to be shaped by human will, then potential outcomes are continuously shifting and changing.

Thus, the modern drive towards innovation and progress* has produced fundamentally different constellations of action, knowledge and responsibility from those arising from traditional responses to the challenges of the future. Contextuality and embeddedness have been displaced by decontextualised, disembedded* relations in order to produce a world of pure potential where anything is possible, thus subject to our design. Having divested the future of content and rooted human freedom in nothingness we find that knowing futures of our making and taking responsibility for them take place under altered conditions: freedom and the committed pursuit of progress are accompanied by an inevitable rise in uncertainty and loss of control. It is here that we encounter the major paradox of the pursuit of progress and the assumption that freedom issues from an open future: *as owners of the future we also carry the sole responsibility for the outcomes of our future-creating actions. This makes us inescapably responsible for that which we cannot know.*

One way to deal with the openness and uncertainty of the future has been to create social rules and regulations that bound and delimit the production of what is unbounded and interactively open. This entailed creating laws and regulations on the one hand and institutions that enforce social rules on the other. In addition, knowledge systems emerged that predicted the uncertain future with new and innovative methods, shifting focus from individual and unique outcomes to aggregate phenomena. The distinction between *facta* and *futura** was established in the knowledge spheres of science and economics where pronouncements were consequently based not on future 'fact' but probability*.

A second way of dealing with the uncertainty of the open future was to treat the temporal realm as if it was spatial and material. This meant approaching it technologically with the tools of material fabrication. As we showed in Futures Traversed, this involved bracketing the open, interactive and transactional aspects of the processes involved. When processes and futurity were thus placed outside the frame of reference and concern, highly paradoxical situations ensued: much of what was externalised emerged from the shadows and dramatically increased the unintended effects of carefully planned strategies and actions. In this modern future-making context it was no longer possible to retain the triple constellation of action, knowledge and ethics in its unity. The interdependencies were severed in line with a growing differentiation of knowledge spheres and academic disciplines whose objects were studied in abstraction from their contexts: the three key elements of the social relation to the future drifted apart. See Figure 3.

The result is a contemporary situation where actions extend over ever longer time spans into the future whilst the sphere of knowledge is reduced to the past and extended present. Since ethics is tied to

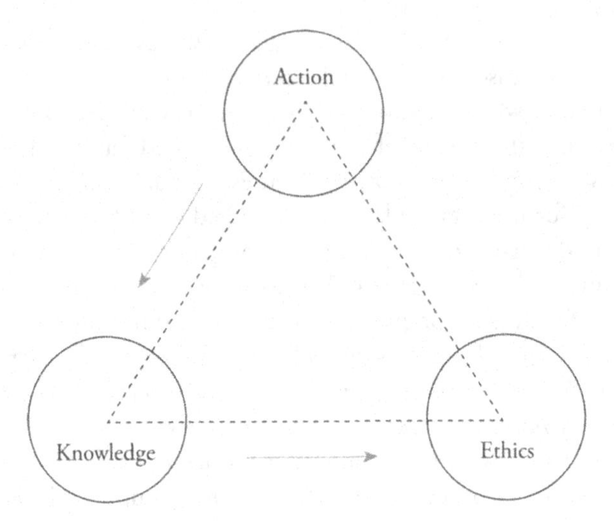

Figure 3: Broken Interdependencies

knowledge of outcomes, responsibility too is restricted to the extended present. On the basis of the taken-for-granted understanding of this relation, environmental economist Michael Jacobs argues that, given we cannot know the future, we should encompass a perspective and duty of care that extends to one generation hence.

> As long as each generation looks after the next (say over a period of 50 years) each succeeding generation will be taken care of. Of course, if an effect in the further future can be foreseen, then it too can be taken into account. (Jacobs 1991: 73)

We can see here a number of broken interdependencies that are of interest to this discussion. First, there is no sense that action, knowledge and ethics form an inseparable unity. Secondly, irrespective of how long into the future the effects of our actions may last, Jacobs implies that we cannot be expected to take them into account if we cannot know about them. Thirdly, when responsibility is tied to knowledge in contexts where the future is seen as open potential, and thus as unknowable, such purely arbitrary cut-off points are encouraged for the suggested horizon of socio-political concern. And since, fourthly, these proposed periods *are* arbitrary and disconnected from their timeprint, that is, their future-making actions and associated consequences, they can be debated *ad infinitum* without ever coming to an agreement. It means, finally, that we are at liberty to continue producing futures without social and ethical bounding, with no one to hold us to account for the long-term consequences of our decisions, actions and inactions.

In the light of such situations, Jay Griffiths (1999: 227) notes critically that the contemporary future of industrial societies "is a blank absence of elsewhere: there is a Teflon coating between today and tomorrow. It is an attitude so implicit it is all but invisible and one merely masked by forecasts, plans and futurism". As no-man's-land, we argued in Futures Traversed, the future is approached as a realm where poisons can be deposited for thousands of years, where resources evolved over millennia can be used up or depleted in a single life time, and where our atmosphere and stratosphere can be altered. Those affected by our future making and future taking have no means of redress. Without voice or vote they are simply at the receiving end of our ignorance. This surely cannot and must not continue. Finding ways to reconnect action, knowledge and ethics is a pre-condition to being able to accompany our actions to their time-space distantiated* effects and to take responsibility in ways that are appropriate to our socio-environmental timeprint.

Hannah Arendt's work is once more of relevance to the issues we are addressing here. Human affairs, Arendt (1998/1958: 183–4) insists, exist in ongoing webs of mutually affecting relationships. As such they are unbounded in space and time which makes their effects unpredictable in principle. As we showed in previous chapters, the bounding of action has been achieved culturally through the creation of social rules and laws, rituals and institutions. Unpredictability, in contrast had been reigned in and stabilised through the production of artefacts. However, the perspective of fabrication has turned out to be no panacea for the unpredictability of social action and socio-technical practices since action ineluctably expands into its networked consequences, and thus cannot be abstracted from its context or reduced to a single deed. When this is attempted, problems occur, as we showed in Futures Transformed and Futures Traversed. Arendt identifies *promise* and *forgiveness* as social responses to, respectively, the unboundedness and uncertainty of actions and their irreversible and unknowable impacts.

In Futures Tamed we showed that the *power of promise* works on the basis of bringing the future into the present. This entails, as Arendt (1998/1958: 245) explains, disposing "of the future as though it were the present, that is, the enormous and truly miraculous enlargement of the very dimension in which power can be effective". The creation of an extended present based on promise, as we have shown, is of course very different from the one that arises with the economic trading* of futures. While promise works in contexts where it can be relied upon, the economic treatment of the future, in conjunction with the valorisation of speed,* negates the very processes that were set up on the basis of covenant. Moreover, we identified discord between private and public modes of being. Our private lives, we suggested, are embedded in unbroken chains of obligation and care that allow us to identify with future generations in a way that is more difficult to achieve in the public domain of instrumental (largely economic) relations where those chains are broken and the frontier spirit* dominates.

Forgiveness, Arendt's strategy to counterbalance the irreversible consequences of actions, works in a very different way and on the basis of unrelated principles. While promise brings the future into an expanded present, recognition of our dependence on forgiveness has the potential to place us in the *future present** of others. It allows us to connect open-ended outcomes of deeds to their eventual impacts on the lives of unknown successors. For Arendt it links us to the planned futures of predecessors through their impacts on our lives today.

> Forgiving serves to undo the deeds of the past [...] Without being forgiven, released from the consequences for what we have done, our capacity to act would, as it were, be confined to one single deed from which we could never recover. (Arendt 1998/1958: 237)

Arendt thus identifies the future-binding covenant and the importance of forgiveness for deeds of predecessors as two social tools to tame the unboundedness, irreversibility and unpredictability that arise with the freedom of social action. In the contemporary industrial way of life, however, neither is still adhered to in the sense that Arendt elaborated for classical antiquity. Yet, both strategies can provide us with pertinent food for thought.

When we apply Arendt's insights to the issues addressed in this book, the need for contexts of social stability where promises can be relied upon is self-evident, if not easily achieved, when progress, speed and economic profit are pursued with such vehemence. The need for forgiveness, Arendt's second means of redemption, is less obvious. Arendt's focus on forgiveness is past-oriented. We forgive predecessors for their deeds and legacies: the cancer-producing radiation, hormone disrupting chemicals, climate-changing fossil fuels. From our futures perspective, however, there is no barrier to extending the temporal orientation of forgiveness to the future. That is to say, we can equally know ourselves to be acting in a context where we not only forgive predecessors but *require forgiveness from successors for our future making*. This turn to the future inescapably embeds us in relations of indebtedness with not-yet existent others, which in turn tempers the frontier spirit and the improvidence with which their presents and futures are spoilt by us and or eliminated altogether. Knowing ourselves to be acting within unbroken webs of obligation, and appreciating that our deeds will require forgiveness, places us almost by default in positions of increased care and concern.

As we identified in our discussion on non-reciprocal* care in the previous chapter, this is the position of, for example, parents or guardians who act in what they think is in the best interest of their children and charges when they choose a particular godparent or a special school. Parents and guardians know that should these turn out to have been bad decisions, they will require their charges' forgiveness for those decisions and the unforeseen problems that arose with them. It would not occur to parents or guardians to insist that they cannot take responsibility for their actions because they could not be sure of the outcomes. Equally, they would not turn solely to past-based scientific knowledge for help to either reduce the uncertainty of the situation or make better decisions. What we have here is an occurrence where *action is directly linked to responsibility in contexts of uncertainty*. It works. We know how to do it at the everyday level, where we are extremely accomplished at linking those two domains of future making. In our daily lives non-knowledge neither incapacitates us nor does it lead to irresponsible action. In contexts of obligation and care, action and responsibility are routinely coupled without giving much thought to the matter. Clearly, what is possible in the private domain ought to be possible also in public life. Important here is first the recognition that responsibility in contexts of uncertainty is possible and second an appreciation of the conditions in which such responsible action can flourish: that is, non-reciprocal, non-instrumental relations of care which are embedded in chains of obligation. Of further relevance is an understanding of the barriers to social relations of care that exist in public life together with strategies designed to break down those obstacles and facilitate the restoration of temporally extended relations of care to the public realm. Much of this has been addressed in the preceding chapters.

The creation of social contexts where promises can be relied upon requires social will. Redemption through forgiveness in contrast is a question of attitudes and values that connect our actions with their effects on future generations. When we approach future making with an attitude that recognises our inevitable indebtedness, it is likely that socio-technical hubris will be tempered. For either of these responses the action-knowledge-ethics relation is held together as a coherent unit but its domains are constituted differently. Where the future can be relied upon due to socially constituted promise, knowledge of outcome is available. In such contexts we can be held to account for our actions. In contexts of non-knowledge, in contrast, this is no longer the case. Here, *knowing ourselves to be dependent on forgiveness from successors for the unintended consequences of our actions becomes important as it induces an approach to the future that is tempered by responsibility to others as yet unborn.* Acting in the knowledge that we require forgiveness is an important step towards relating what has come adrift, acknowledging interdependence and our implication in time-space distantiated effects, and widening our horizons of concern, obligation and responsibility.

A further step has to be achieved through conceptual renovation. This entails not simply the recovery of enfolded wisdom but requires in the first instance that we make visible historically sedimented habits of mind. Some of these ways of knowing have become inappropriate and are thus in urgent need of change. Others, which had been displaced, have become relevant once more. Thus, their recovery and subsequent adaptation to the contemporary context become pertinent. At this point, therefore, we want to reconsider some unquestioned assumptions that have been naturalised as 'truths', bring some of their key features to the fore and explore openings for change. This process is important, we argue, since a new way of understanding future making and the relation between action, knowledge and ethics is a precondition for change at the level of both individual action and public policy.

Throughout this book we have worked with the assumption that knowing is intimately connected to doing, theory to practice, that understanding which is *in*appropriate to the contemporary condition is therefore tied to equally *in*appropriate action. To stress the performative nature of knowledge Marx worked with the concept of *praxis.* We have mostly used the term *knowledge practice* to express this active and constitutive side of knowledge and to convey our conviction that transformed understanding and new knowledge affect our action potential, enhance our capacity for change. The point of troubling taken-for-granted assumptions therefore is to open up spaces for doing things differently. We can here only select a few of the most pertinent assumptions we encountered in our exploration. Where we have dealt with problematic presuppositions in depth in the text already—the 'empty' future, the future as 'spatial territory' and the future as a 'free resource', for example—these will not be revisited here. Instead, we will address the reality status of the future and its moral standing. Pinpointing some of the key inadequacies allows us to open up spaces that enable us once more to consider alternatives appropriate to our context and sphere of influence. Some of the summaries that follow are more black and white than we would wish but sometimes such simplification and contrast are necessary to achieve a measure of (artificial) clarity before complexity is reinstated to its rightful place.

Troubling Facts

In this section we explore the reality status of the future. This takes us into the realm of presuppositions that are deeply embedded in western cultural history and associated with the rise of science to dominant knowledge system. Today, mechanistic science's way of understanding the world has become the unquestioned western norm for evidence-based practice. It is used, defended and legitimated by the governments of industrialised nations. It is deferred to in the media and in courts of national and international law. Its assumptions have become deeply embedded habits of mind that are difficult to unsettle and even harder to displace.[1] The difficulty, however, should not deter us from trying.

When pared down to the bare bones of its temporal logic, mechanistic science deals with facts and its operational domain is the present. These facts refer to past events: done, achieved, completed and thus amenable to empirical investigation. Facts can be established as evidence. The future, in contrast, is that which has not yet come about, something non-factual which *will* become fact only after it has occurred. While the one has already taken (unalterable) form, the other is still open to influence. Moreover, as we showed in Futures Traversed, facts belong to the realms of space and matter. The non-factual future in contrast is associated with mind and the realm of ideas. It is desired, anticipated, expected, planned and projected.

[1] As noted in the Prologue, we are here not discussing the activities and assumptions of contemporary scientists but rather the reductive use of science in everyday public life with its associated taken-for-granted expectations about what science is and can do. Our critique therefore is not of modern science but rather of the unquestioned public use and abuse of mechanistic science for purposes of legitimation. On the role of science in legitimating practice, see for example Haack (2005).

This understanding, which has become an unquestioned 'fact' within the western tradition of thought, was cogently expressed by St Augustine in the fourth century AD.[2] To live life as a human, St Augustine thought, involves the interaction and integration of past memory, present perception and future anticipation. He concluded that only the present exists while past and future are aspects of the mind only. St Augustine's understanding of the past, we need to appreciate, has been partially abandoned and replaced. Thus, today we no longer think of the past as an exclusive domain of memory. Instead, we acknowledge that the past has also left records and traces. It is on this basis that we can know the invisible past as fact. Moreover, our methods of accessing those hidden traces are still developing, so that today, with carbon dating, for example, we can know a 'factual past' that extends over millennia. St Augustine's view on the future, in contrast, has survived unscathed in its original form. Here, there has been no equivalent development for gaining access to 'factual processes in the making', to futurity that extends into the long-term future, which has not yet congealed into phenomena. Today, as we have shown, past, present and future interleave and futures are not merely planned or imagined but set on their way. The assumption and associated distinctions therefore no longer hold. The ancient perspective on the future loses its grip. *After 1600 years of adhering to the non-factual understanding of the future it is time for a change.* Contemporary contexts where past and present futures* are already in progress require that we grasp *as real* latent processes that set future presents in motion.

The important point for this discussion on habits of mind is that currently only outcomes of processes are accorded reality status. Where processes take a long time to congeal and effects cannot be linked causally to an origin, the invisible process domain tends to be negated, placed outside the scientific frame of reference. The mechanistic scientific perspective tends to lack conceptual tools, an appropriate methodology and adequate explanations to grasp time-space distantiated processes in progress. This is a problem since today a great number of the products of science, such as chemical, nuclear, genetic and nano-technologies, for example, are characterised by processes of extensive time-space distantiation, that is, the stretching of effects across time and space where the latency period is vast and effects cannot be linked unambiguously to

[2] See Bourke (1983); for a brief summary see Adam (2004: 52–4).

their causes. We can therefore say that *the relevance of factual understanding that accords reality status only to processes that have congealed into matter is decreasing proportionally to the increase in technologies marked by long periods of latency.* This situation too is clearly in urgent need of change.

How, then, is this deeply ingrained set of assumptions to be troubled and opened up for change? We can, as we have done in this book, amass examples from the breadth of social action that show its inappropriateness for the contemporary condition. It is, however, still a big step from recognizing the inadequacy of trusted sets of assumptions to changing them. The alternative to the *status quo* not only has to be able to do what could be done before but also has to do more and do it better. In other words, it has to be an all-round improvement. Moreover, the concepts chosen as replacements have to resonate with experience at the everyday level and do the job that is required of them.

Let us begin by considering the concept of *process* in the English language. The difficulty can be appreciated when the English notion is compared respectively to a Latin and German pair of concepts that encompass the process world. In Latin we have the distinction between *natura naturata**, the world of factual outcomes and *natura naturans**, the active process world of nature in the making.[3] The former is factual, finished and finite. The latter is temporal, transient and transformative. *Natura naturans* thus refers to the activity and creativity of naturing, to nature in the process of its production. In Germany, Jakob von Uexküll and Georg Kriszat (1983/1934) have worked with the related distinction of *Merkwelt**, which is the factual world amenable to perception, and *Wirkwelt**, which is the active, creative and productive world of processes. The *Wirkwelt* is largely invisible. It is marked by projective interiority and depth. It is oriented, yet without simple or fixed location. Its formative activity is below the surface, inaccessible to the senses and thus beyond the grasp of empirical science. This means that its activities and processes need to be intuited and unfolded. Inaccessibility to factual investigation, however, does not diminish the importance of the *Wirkwelt*, its productivity or its reality status. Rather, all the power of activity and production of impacts belongs to the *Wirkwelt* rather than the visible world of outcomes. The *Wirkwelt* is the domain of lived*

---

[3] Baruch Spinoza, in his *Ethics* (1992/1677) is generally credited with having drawn attention to this important distinction.

and living* futures, discussed in the previous two chapters, while the *Merkwelt* signifies the finished product, the process-world congealed into material form, surface phenomena occupying space and amenable to quantification.

Throughout the previous chapters we have argued that past and present facts are the bounded products and congealed form of processes, nature in its phenomenal form, snapshots of the ephemeral world of change. This means we have worked with the distinction between *products* or outcomes of processes (past and present) and *processes* that produce outcomes (futures). The difficulty is that in the English language the concept of process lacks some of the power of both the Latin and the German characterisations respectively, which stress the active, creative and transformative character of the process domain, thus emphasise its futurity. This deficit in the English concept of process makes it that much harder to grasp the reality of futurity with reference to invisible and latent processes which may not materialise as symptoms for a very long time.[4] To our knowledge there is no ready-made equivalent pairing in the English language. In order to achieve the distinction between accessible, temporally bounded outcome and inaccessible, temporally unbounded, transformative futurity we therefore would like to propose *phenomenal reality** for *Merkwelt* and *effecting reality* for *Wirkwelt.*[5]

Gilles Deleuze and Félix Guattari's (e.g. Deleuze 1994/1968) theorization of the *virtual* and its 'halo' of potential, discussed in Futures Thought, does much of the work we feel is necessary to accord reality status to the realm beyond empirical grasp. Their use of 'the virtual' as the key concept for this re-conceptualisation, however, is unfortunate as the term itself has many inappropriate terminological associations. Through its everyday usage—for example, as 'virtual reality' in the world of computers—it is infused with *un*reality and even the best theoretical argument will not overcome this problem of association. The second, more profound difficulty with the concept of the virtual is its a-temporality

[4] For an extended discussion on these matters see Adam (1998) where the distinctions are first discussed on pages 33–35 and then applied to socio-environmental phenomena throughout the book.

[5] In our search for appropriate English terms we consulted colleagues and would like to express our thanks to them here: Dr Rachel Hurdley for the helpful excursions into Latin grammar to better illuminate the distinctions between *natura naturata* and *natura naturans* as well as *facta* and *futura*; Dr Jan Adam for our in-depth discussions of the German terms *Merkwelt* and *Wirkwelt* that led us to the English distinction between 'phenomenal reality' and 'effecting reality'.

in everyday usage, which provides us not even with a hint of process, let alone futurity or creative, effecting power. It is static and decontextualised, its location everywhere and nowhere. In conjunction with the simulation of animation heroes fighting it out in the world of computer games, the virtual becomes unusable as an alternative to the assumption of a factual present and a mind-based, thus fictional, future. Finally, in everyday usage the opposite of virtual is real. Since every concept is co-defined by its other, the everyday opposition between the real and virtual makes the virtual unreal by default. On these three counts, therefore, we find the concept of the virtual unhelpful for the task of grasping as real the effecting, processual future in progress. This means we need to take on board Deleuze and Guattari's important theory of futurity (expanded by our conceptualisation of lived and living futures), retain its reality status but abandon the terminology of virtuality for its problematic everyday imagery and inappropriate associations.

In this book we have developed the idea that the future is both *lived* and *living*. It is lived, we argued, at every level of reality. As futurity it is lived by humans as social beings, always ahead of themselves, extending to what they and others will be and become, to the horizon of individual and collective death as well as the legacies they leave behind, which will grant them immortality. As futurity the future is lived also at the level of organised organic and inorganic matter, albeit in degrees of receding consciousness. Even matter such as sand and stones are ahead of themselves, extending temporally and interactively to future states: in conjunction with water and wind, for example, the stone will turn to sand and, depending on context, the sand may turn to sediment at the bottom of the ocean or form dunes and grow plants where its nutrients live on in chains of other life forms. Context dependence makes this future like all others both predictable *and* unknowable. Futurity is 'Being ahead of itself', to use Heideggerian language, and potential for novelty. It encapsulates the inescapable reaching out from an ever-changing present.

The idea of the living future, in contrast, has a different location within the time-space-matter continuum. It is to be found in the interactions, the patterns, processes and rhythms of change and evolution beyond individual perception. It is rooted in the wider ecological give-and-take that extends from the beginning to the end of time within which our interactions and socio-technical products are embedded and where we partake as participants and contributors across the levels of being. It designates a continuum of variable pasts and futures, extending in unbroken chains of interactions to origin and destiny. It is the basis on which we can know ourselves as star matter. The idea of the living future, therefore, can offer the crucial active and creative ingredients to the effecting world of futures in the making which is entailed in the German *Wirkwelt* but lacking in the a-temporal English concept of process.

Finally, we have distinguished *present futures* from *future presents**. The present future refers to the standpoint of the present. As such it encapsulates both the factual approaches of science and economics and process-based perspectives on futurity and the lived future. The concept of future presents, in contrast, encompasses the future as both an effecting process and/or as living. Moreover, the *standpoint of the future present positions us with reference to potential impacts of present actions on future generations who have to cope with the consequences of our inventions and interventions.* That is, it relates us to deeds and processes already on the way.[6] Through their different positioning, present futures and future presents offer greatly divergent options not only for knowledge but also for action and ethics. Only the standpoint of the future present, we need to appreciate, enables us to accompany our actions to their potential destinies and know ourselves as responsible for their time-space distantiated impacts. To encompass future presents and to take that standpoint, however, requires that we first understand as real and living these invisible, effecting process futures in progress. This prior move is essential if we seek an approach to the future that brings into a coherent unity action, knowledge and ethics. While we have moved quite a way in that direction already, there are, however, still a number of hurdles to overcome. Prominent amongst these is the widespread practice of reductionism.

Complexity with Futurity

Looking across the thought traditions of western cultures, we can recognise that reductionism in its various guises has facilitated socio-technological development and control. In fact it has been central to the successes of the industrial way of life as well as its excesses. Admittedly, in this book

[6] As we have explained previously and show in the Glossary, Niklas Luhmann (1982) based the distinction on the difference between mentally represented utopias, which open up present futures, and the technology-based reality of future presents, which close down options for both present and successor generations.

we have concentrated on showing interdependences primarily with regard to the excesses rather than the successes of this way of life. Thus, in a number of previous chapters we have shown how reductionist approaches to reality have been implicated in the fragmentation of social existence and knowledge on the one hand and pose significant barriers to a futures perspective that seeks to re-unite action, knowledge and ethics on the other. The shift from multiplicity and interactive complexity to abstracted simplicity, from the interplay of exteriority and interiority to surface phenomena, from the complex interpenetration and mutual implication of time, space and matter to matter in space are just some of the reductions we encountered in this exploration.

In today's world of rising complexity and interdependence this kind of reductionism is being questioned. Across the knowledge spheres from physics to philosophy emerge discussions about the need to embrace complexity, multiplicity, context and the temporal world of process and change.[7] These efforts are at advanced stages of development and writings on the subject have begun to proliferate from the last part of the twentieth century onwards. The detailed debates and distinctions are not at issue here. What is of importance, instead, is the potential of this sweep of conceptual changes for transcending contemporary relations to the future. Since the complexity perspective across disciplines is hailed as the solution to reductionism, we need to establish whether or not it helps us to reinstate the active, creative and effecting process domain of futurity. Does it enable us to accord reality status to futures in the making, we need to ask, no matter how vast their time-space distantiation?

Fritjof Capra (2003), in a key text on complexity theory across the knowledge domains, proposes that the social dimensions of matter, space, time and meaning need to be brought into a coherent relation. Material structure, spatial patterns of networked relations, temporal processes of becoming and the cultural meanings these hold, he argues, need to be given equal weight in our analyses. The spheres of matter, space, time and knowledge have to be seen as mutually implicating rather than mutually exclusive. Moreover, when they are integrated in one analytical framework, context becomes an important consideration. As such, the

[7] See for example Beinhocker (2006), Bohm (1983), Briggs and Peat (1989), Byrne (1998), Capra (1996), Gribbin (2005), Hayles (1990), Law and Mol (2002), Luhmann (1982), Nowotny (2001), Prigogine & Stengers (1984), Thrift (1999), Urry (2003), Weick and Sutcliffe (2001), Wynne (2005).

complexity perspective unsettles tried and trusted schemas for coping with the unknown. It questions the perspective that has habituated us to expect certainty, depend on simplicity and trust past-based evidence. Importantly for our discussion here, it allows for changes to the way the future has been handled for the past three hundred years, that is, to the historically tempered deep structure of cultural engagement with the not yet and the unknown.[8]

In Capra's fourfold constellation of complexity theory each aspect implicates all the others. When, in addition, we infuse Capra's perspective with futurity, as developed in this text, then new possibilities for understanding open up and the frozen world of facts springs into life. A brief summary of Capra's complexity constellation, which we have extended to encompass temporality, shows the potential of complexity theory for knowledge practices that seek to embrace futurity.

Matter, Capra's first domain, is our physical world—the earth we live on, the soil that feeds us, the air we breathe, the water we depend on, the body we inhabit, the landscapes and cityscapes we dwell in, the other beings we co-evolved and co-exist with and the socio-cultural world of artefacts (buildings, books, tools, machines, vehicles, computers, power stations, and laboratory products). From a futures perspective, however, this matter is to be understood not just spatially as frozen in time but also temporally as extended and enduring, interacting and regenerating, decaying and leaving a record, projecting and entailing *for*-ness, that is, futurity.

Form, Capra's second part of his fourfold constellation, encompasses patterned and networked relations of family and friends, work and play, with domesticated and non-domesticated other species. It covers all infrastructural aspects of social life, such as institutions and communication systems as well as political, economic, religious and knowledge-based associations. From a futures perspective this form is to be expanded. It is to be grasped not only synchronically as structured pattern but also diachronically: form as forming and historically formed, network as networking, pattern as patterning.

Process, Capra's time dimension of complexity, focuses on the temporal aspects of the world of space, matter and networks. It relates to the way this world is produced and to emergent properties arising from

[8] For a working paper that elaborates on the temporal perspective on complexity, see Adam (2005).

interactions. From a futures perspective this world to has to be extended to further encompasses 'for-ness', extension into the future, futurity. It needs to include understanding of the dynamics of change and creativity, stability and novelty, continuity and discontinuity, evolution and history, a dynamic that produces not just emergent presents but entire *futurescapes* of past futures and future pasts, present futures and future presents, processes and their products as well as lived and living futures. It has to acknowledge a world where much of the on-goings and their effects are stretched across time and space, therefore often latent and invisible until they materialise as symptom—sometime, somewhere.

Meaning, the fourth feature of Capra's complexity perspective, involves the processes and products of reflective consciousness as well as socially constituted knowledge such as language, values and beliefs, which tend to be tied to the present or the a-temporal realm of ideas. From a futures perspective, however, meaning is projective and action-oriented. Knowledge is performative and transformative, hence we prefer to use the concept of *knowledge practices* to that of meaning. In their temporalised form, knowledge practices resonate with process and becoming, with form as historical and projective forming. This understanding in turn needs to acknowledge the contextuality of meaning and recognise knowledge practices as embedded and interdependent with the entirety of our world thus not abstractable from their networked relations. It places each one of us in the position of responsible social agent and future maker and thereby leaves behind the 'view from nowhere' that allowed us to act with impunity. Heidegger's *Dasein**, as discussed in Futures Thought and Futures Tended, is to be conceived no longer as merely individual but also social and collective.

From the above we can see that the complexity perspective requires non-linear thinking, that is, understanding of networked interdependencies and processes in a reflexive, autopoietic, non-sequential, non-linear way. In its non-temporalised form, however, there remain some major obstacles to utilising the potential of complexity theory for a futures perspective. Central amongst these is the way linear causality has been retained without the necessary adaptation to the requirements of the complexity viewpoint. This is particularly troublesome from a futures standpoint. As we have indicated in earlier chapters, our understanding of causality is linear, sequential, reductive and past-based, which has significant consequences for our concern with futurity, future presents and the process world of living futures.

In Futures Thought, we noted that Aristotle conceptualised *aition*, which is generally translated as causality, with reference to four interdependent elements. From our futures perspective these four elements appear to perfectly parallel the quadruple complexity proposed above, as long as meaning is temporalised and thus conceived as transformative knowledge practice:

Aristotle's causes	*Complexity dimensions*
Material cause	Matter
Formal cause	Form
Effective cause	Process (past-based)
Final cause	Meaning (projective knowledge practices), futurity

In the course of their historical development the natural sciences have reduced Aristotle's first three causes to one general physical cause where action produces subsequent effects in a linear fashion from past to present and future. The idea of a 'final cause' as both for-ness and the goal or end towards which organisms develop has been eliminated altogether. With this simplifying move the temporal has been reigned in and futurity effectively shielded out from scientific causality. In its place, the past and the a-temporal present have been installed as exclusive sources of the scientific meaning of causality.

From the above we can see that the complexity perspective provides us with the potential to transcend that reductionist, linear, a-temporal understanding and take account instead of the complex, interdependent, temporally extended social realm of matter, relations, processes and knowledge practices that produce time-space distantiated material effects. It is because our knowledge practices have impacts, which extend materially, spatially and temporally, that we need to explore ways that allow us to accompany the consequences to their eventual, potential destinies: tomorrow, in one hundred, even one thousand years' time. For this we need a causal understanding that transcends a mechanistic science perspective. The new complexity conceptualisation of causality needs to achieve the following: first, it has to implicate each of the other dimensions in any one aspect explicated. Secondly, it has to extend not just from past to present and future but also from future to present and past. Thirdly, this future-to-present direction needs to be not merely an aspect of mind, that is, of our imagination, but also a materially constituted effecting reality. We need to know it as for-ness and as deeds under way, as lived and living futures, as futures in the making that cast shadows from the present to the future and back again, not yet congealed into matter but material nevertheless. And this is precisely the point: future-oriented and future-creating knowledge practices produce living futures that reverberate through the entire system of physical, biological and cultural relations and processes. Aristotle's four causes, therefore, provide us with a base from which to start our reconceptualisation of causality in a way that is consistent with the futurity and temporal complexity we seek to encompass in our understanding. Thus, we can say, when complexity is complemented by temporality and futurity then the active, creative and transformative domain of futurity is encompassed in the understanding and futures in the making become visible in their effecting materiality.

With the four-fold understanding of complexity and causality we can appreciate what was inaccessible before. Our knowing becomes reflexive. When we grasp that knowledge practices are neither isolated nor isolatable from their networked connections, that our deeds reverberate through the system, activating responses that stretch across time and space and are therefore not necessarily proportional to their initial 'cause', then we are also bound to recognise that we are implicated participants that cannot escape their responsibility. The complexity perspective deprives us of the comfortable position of external, uninvolved observer. It divests us of the 'view from nowhere' that allowed us to act with impunity. It therefore demands that we acknowledge ourselves as future makers and understand our responsibilities accordingly. Here too, however, we find that deeply engrained habits of mind stand in the way of taking that responsibility seriously and prevent us from relating action, knowledge and ethics in a meaningful and coherent way.

Beyond Certainty

When we shift emphasis from assumptions associated with knowledge of the future to assumptions about the responsibility part of the action-knowledge-ethics relation, as we have done in the previous chapter, the first thing to note is that both legally and morally we feel exonerated from responsibility when outcomes could not be foreseen at the time of action. With respect to the nuclear industry, for example, we find that the people who counselled governments on whether or not to establish a nuclear capability, and who happened not to include in their considerations associated problems of safety, were and still are not being held legally responsible for either the resulting health hazards or the economic burden of the billions of dollars required for the decommissioning of power plants and the management of radio-active waste. In the nuclear case we find that installations are covered by limited liability only, which means that society is expected to foot any bill that might arise with accidents or leakages. This explicit recognition of the (non)knowledge-responsibility link is even enshrined in law. Thus, the Price-Anderson Act was introduced in the USA in 1957 specifically to limit the liability of nuclear power plant operators in the event of an accident.[9] It has been renewed several times since, and now limits the amount of liability for each site to $300 million. Thus, whether formally or informally, non-knowledge as well as unintended and unforeseen consequences all absolve us from personal and public responsibility.

Yet, for some socio-technological unforeseen effects the tide is turning. Thalidomide, asbestosis, smoking-related diseases and similar technologically produced hazards are cases in point where companies are being held responsible for the harm produced by their products. Thus far, however, such apportioning of responsibility for time-space distantiated effects applies predominantly to cases where causal chains can be established over the life times of individuals. It is not clear as yet, what happens to responsibility in situations where effects do not materialise as symptoms for hundreds and even thousand of years. The contemporary problem is that we link responsibility to knowledge in contexts where increasingly *non*-knowledge is becoming the dominant feature and thereby create ever-increasing spheres of *ir*responsibility. This is an unsustainable situation in desperate need of change. Since, however, the underpinning assumptions reach back in western cultural history to Greek antiquity, a change in these deeply embedded habits of mind is complex and involves not one but a number of suppositions and beliefs. In Futures Tended we have built on Hans Jonas' (1984/1976) work, locating these beliefs and their impacts in the wider cultural setting. Here we want to briefly highlight them, show their inappropriateness for the contemporary condition and identify some of the key features that would need to change for these moral presuppositions to become appropriate to contemporary future-making practices.

In models of morality with roots in Greek antiquity, responsibility is generally thought of as pertaining to relationships between living

[9] See Shrader-Frechette (1993: ch. 2).

individuals. Actions involving non-human things such as artefacts and technological products, in contrast, were not considered of ethical significance. Moreover, virtuous moral action was to be achieved in the here-and-now world of politics. This meant that moral action and matters of ethics were defined by close proximity, thus limited in time and space. The long-term future, in contrast, was associated with fate, providence* and destiny. It was the realm of gods, and was not subject to human planning, debate and moral action. As such it was outside the sphere of human responsibility. This present-based morality was counterbalanced by an ethical orientation to eternity, regarding the good and the beautiful, truth and virtue, ideas and ideals. *Responsibility of individuals and political leaders was consequently defined by eternal values, which were to be enacted in the present by members of particular communities.*

In contrast to the Greek model, obligation towards a technologically produced, long-term future arises with the age of science. It emerges first with the capacity to create futures that outlast their originators, secondly with the human potential to threaten not just individual existences but the continuity of our species and life as we know it, and thirdly with the pursuit of progress which destabilizes eternal values and renders them historical. This context for responsibility is new. Today, the foundations for responsibility have shifted from an exclusively individual to a collective base, from predominantly local to global effects and from primarily present impacts to actions that may not materialise as symptoms for a very long time. The common-sense ethical assumptions, which we have inherited from the Greeks, therefore no longer hold for the contemporary condition. Let us explain by once more using nuclear technology as our example.

Beyond Immediacy: the effects of today's socio-technical, socio-economic and political processes are no longer spatially or temporally bounded, this is nowhere more pertinent than in the case of nuclear technology. Radiation, although most dangerous in the immediate vicinity of any leakage or accident, permeates outwards in space, spreads inwards in matter, organisms and bodies and extends temporally into the long-term future. Moral principles grounded in the immediacy of the here and now, therefore, need to be adjusted to the timeprint of potential outcomes. Such expansion of responsibility to the potential reach of actions places us in a different position with respect to what can and cannot be known, done and controlled. This means that responsibility can no longer be routed via knowledge. In contexts of extensive time-space distantiation (and the associated predominance of non-knowledge),

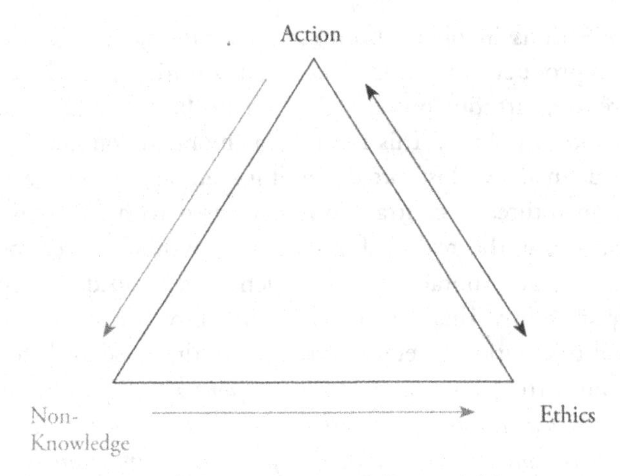

Figure 4: Uncertainty and Future Making

therefore, a direct link has to be established between action and ethics, ethics and action. See Figure 4.

Beyond Individual Responsibility: through the ages responsibility had been associated with individuals and their deeds. While this still holds good today, especially in the application of our laws, for example, technological activity in general and the policies associated with nuclear power in particular have the potential to affect the living conditions of *all* people now and in the future. This is not to suggest that the impact of decisions regarding radioactive waste management, for example, will be equal across time and space, but simply to point out that the time-space dispersal of effects is no longer encompassed by a moral code focused on the actions of individuals. The changed context means that the ethical project of modernity has to be expanded beyond individual responsibility to encompass collectives at the national and international level. Thus, for example, in recognition of the trans-boundary nature of radiation, nuclear policies have to become a cosmopolitan endeavour. Moreover, since liability for associated costs continues to be increasingly externalised to society at large, it is society who needs to come to decisions about the nuclear present and future. Not politicians whose mandate expires after their period of office, not scientists who build and maintain the installations, not insurance companies who cover limited liability but the general public who are liable will have to debate the pros and cons of that for which they are held responsible and for which they will require forgiveness from successors if their decisions lead to disasters sometime, somewhere.

Beyond Anthropocentrism: the transformative power of humans has always been extensive. In the industrial age, however, this capacity has reached undreamt of heights and fundamentally changed our relationship to nature. Today, nature is no longer the mere backdrop to human action but is subject to scientific intervention and invention. Flora and fauna, mountains and valleys, riverbeds and oceans, the biosphere and atmosphere—all are influenced by scientific practice and its technological applications. As such, nature in all its facets has become ethically significant, without, however, having its 'interests' represented in the socio-environmental polity of today. Instead, human interests grounded in the short-term politics of the here and now, arbitrated by science and justified on the basis of economic arguments are the primary determinants for decisions that impact on the long-term future of our environment and fellow beings. In the light of this mismatch between ethical assumptions and the reach of socio-technical effects we are charged to rethink our traditional anthropocentric responses and produce principles more appropriate to our ecological footprint and timeprint. This requires opening up ethical concern to encompass, as our responsibility, the sphere of impact, which extends beyond humanity to all of nature and the physical bases of our existence.

Beyond Certainty and Control: while the future has always been uncertain, humans were not called upon to take responsibility for what was considered the realm of gods or God. They were merely required to act responsibly in and towards the realm that did not belong to them. In a secular social world, which is understood to be (to a large extent at least) the outcome of human action, in contrast, the *unknown* and *unknowable* futures of our making become our responsibility. That is to say, uncertainty of potential outcomes cannot absolve producers of long-term, open-ended impacts from responsibility to those affected in remote futures and places. The difficulty confronting us, as we have shown, is that the *indeterminacy of unbounded effects makes reliance on scientific prediction and economic risk calculation inappropriate and presents us instead with questions about justice, rights and possible harm to future beings that have to be addressed.* Despite the extensive scale of potential effects and inevitable uncertainty we need to accept therefore that responsibility extends to the reach of our actions. This principle applies irrespective of whether or not the affected and afflicted are able to hold us to account. However, once we accept this general principle, as we have argued above, we need to find ways of connecting action to responsibility without routing it via knowledge. For a timescale of

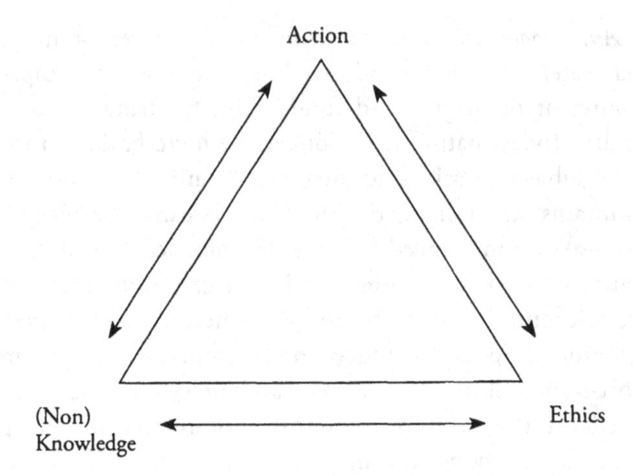

Figure 5: Reconnecting Interdependencies

action where outcomes can no longer be tied reliably to their sources and where knowledge becomes inescapably fuzzy, therefore, it is not past-based knowledge but *social* debate on what is right and just which will need to arbitrate between competing plans, decisions and interests. When we accept, in addition, that decisions may require forgiveness from successors, we act in a social context of indebtedness and this in turn helps to temper economic rationality and present-oriented self interest. It means that relations between action, knowledge and ethics need to be organised flexibly rather than uni-directionally. See Figure 5. How the interdependence is conceived and socially constituted, however, will depend on specific contexts of action and their attendant *timeprints*.

Reflections

In this book we have placed contemporary approaches to the future in a wider historical frame in order to give us a base from which to make comparisons, to identify differences and appreciate continuities. From this expanded perspective we could begin to understand what had been gained and what lost on the path to modernity. We could see some of the impacts associated with the major shift that has occurred in the ownership of the future, that is, people taking charge of the temporal domain that had previously been the preserve of gods. We could recognise a number of paradoxes that have accompanied the results of this transformation and appreciate some of the underpinning interdependencies. In the course of reading the book possible openings for change became tangible and in the previous two chapters these were identifiable at the level of theory and ethics. In this chapter the potential for doing things differently became apparent at the level of implicit assumptions and naturalised habits of mind. Transcendence of contemporary approaches to the future becomes a real possibility once action, knowledge and ethics are reconnected and placed in relation to each other. Knowing that things could and can be different empowers us to infuse future making with concern and responsibility appropriate to our timeprint.

EPILOGUE

What a journey it has been: a conceptual expedition into extraordinary territory. At the end of it we find ourselves poised between origin and destiny, at home between the past and future. An explorer's life is often arduous, and during the most difficult times it requires fortitude and trust in both the project and one's capacity to achieve the exploration's objective. Above all, however, it is a life of immense privilege. Whether an explorer discovers virgin territory or recovers lost lands, each find and every insight lifts the spirit, enthuses and enchants. Looking back, of course, the new always looks so familiar, so utterly sensible and mundane. Looking back all you can think is 'yes of course, why did we not see this before? It all makes perfect sense. Surely we knew this all along'? And so, in a way, we did. We did, because our exploration has tapped into a knowledge base, encoded in nature and culture, that constitutes who we are, what we know and, to a large extent, what we believe in and what we do. It made the familiar (the taken-for-granted common sense) strange, rendered invisible interiority tangible and (re)connected us to our temporally distant selves, both past and future.

All along the way we were carried by the confirmation and the associated strength of feeling that, yes, there had been other ways, there are different modes of being, the industrial way of life is not destiny. The valorisation of speed, the tunnelling pursuit of profit, the autistic present-orientation are just phases of cultural history and mere blips in the evolutionary scheme of things where futurity has been, is and will be handled differently. More encouragingly still, even after three hundred years of industrial societies developing this logic of 'present nowhere', people's daily lives among family and friends are still conducted to a temporal logic where action, knowledge and ethics are held together with great skill. Futurity is lived in a context of living futures and pasts that matter.

Our journey has thus been one of many r's: recovery and restoration, renovation and re-appropriation, relation and re-connection, restitution and redemption and above all re-enchantment. We recovered the existential challenge presented by our futurity and encountered ancient and modern ways of dealing (or not dealing) with it: change and the ensuing uncertainty, transience and the accompanying impermanence, mortality and finitude which accompany every moment of existence. The evolutionary ways of dealing with these challenges will continue to develop. The cultural responses come and go, but it has to be admitted that some were more stunningly successful than others. The strategies associated with the industrial way of life are at best in a phase of transition, at worst singularly useless as responses to the existential challenges that pervade our being at the level of both culture and the material body.

We were involved in the deeply satisfying work of restoring the temporal realm to the domains of space and matter and experienced with great joy the transformations, traversals and transcendences that ensued. We sought to hone our thinking tools to fit their purpose of contemporary future making in a mode of concern and care that is cognisant of our timeprint. This meant that some conceptual tools were in need of renovation others required re-appropriation and reclamation for new purposes. Here, some of the important groundwork is completed. Change, however, requires that the tools are used. It necessitates practice and this needs to be of a particular kind: not fragmenting, not abstracting, not decontextualising but relating and reconnecting, embedding, embracing and embodying. It asks that we recognize ourselves as future makers whose actions reverberate not just in and through our circle of familiars and colleagues but through the lives of others distant in space, time and matter. Seeing those connections and interdependencies became a primary task for this exploration. How wide the web will be spun, how well the threads will be connected, how effectively action will be related to knowledge and ethics depends not just on what we have written but more importantly on the responses and actions those words have provoked.

Finally, there were the staging posts of restitution, redemption and re-enchantment. We did not deal with these explicitly, never treated them as separate issues to be discussed. Consequently, they are to be found between the lines, the sentences and the pages. Restitution and redemption were motivating us and keeping us on the path when the going got particularly tough, when fog rather than clarity prevailed, when we no longer could see any way ahead. The future does not belong to us. Something needs to be done to restore it to its rightful owners. If we do borrow from our successors, as we must, then what are we giving them in return to thank them for their generosity? Radiation, hormone-disrupting chemicals, and genetically modified organisms seem somewhat inappropriate. We sought redemption. In small measures and tiny steps we were making headway in that direction. Whenever we managed another little step, arrived at a clearing, achieved new insights and far-sights, the joy was immense. In those moments we felt back in touch with our soul and spirit. Our lives were re-enchanted. Knowing ourselves poised between origin and destiny, we are at home between past and future, especially the future.

Wanted: 21st Century Experts on the Future

Experts are sought to restructure contemporary institutions in ways that take account of long-term effects of scientific developments and their contemporary socio-technical applications.

Requirements:

- Competence in the fields of future-oriented action, knowledge and ethics
- Historical understanding of differences and continuities in social relations of the future
- Understanding process worlds and their invisible, time-space distantiated effects across all levels of being
- Appreciating the difference between present futures and future presents and being able to do justice to both
- Knowing where the public domains of science, economics and politics end and where responsibility is inescapable at both the individual and collective level
- Recognizing themselves as objective but inescapably *implicated* participants, thus combining objectivity with normativity
- Developing appropriate theories and methodologies
- Expertise in the policy world so that the configuration of future-based action, knowledge and ethics can be flexibly applied
- Enthusiasm to inspire others to follow on what initially will be a lonely path.

GLOSSARY OF KEY TERMS

Words italicised in the text below refer to other glossary entries. When a glossary term appears in a chapter for the first time, it has been marked in the text with an asterisk (*).

Abstract future

The future interpreted as a predictable product of the past. This is the future as known to classical Newtonian science, in which mathematical laws based on past observations accurately describe the future position of bodies. It is the result of regular combinations of natural events governed by unchanging laws.

Commodification of the future

Treats the future as an economic resource where the potential benefit of an action is calculated in terms of its expected return for the present. The future is fully commodified once economic reasoning about profit and loss has become the dominant social form into which ideas of progress are translated. See also *Discounting the future* and *Empty future.*

Care

The everyday term, which inescapably encompasses the future, takes on specific significance in the philosophical thought of Martin Heidegger (1988/1927) where it forms the basic character of all human dealings with the world. Nothing is encountered without it 'mattering' to us in terms of our 'projects' which extend us temporally beyond experience of an ephemeral present.

Conatus

Latin term used in medieval philosophy to refer to a basic drive towards self-preservation inhering in living things, having its root in *conatur*, 'to endeavour'. It is used by Hans Jonas (1982/1968) to describe the fundamental involvement of all life forms in their own futures and stipulated to operate from the level of cells upwards. See also *Lived future.*

Constitutive Value

Is being valuable as a part of something else, but only insofar as it is allowed to exist in its own right and for its own sake (e.g. the value of people, things, institutions and ideas insofar as they contribute to the flourishing of networked relationships across space and through time). It is to be contrasted with intrinsic value (the value something has in itself, independent of its relation to other things) and instrumental value (something valuable for the sake of obtaining something else).

Dasein

Heidegger's (1988/1927) term for 'human being', meaning literally 'there-being'. It entails that we experience the world as unfinished, as always requiring something more from us, thus we are always ahead of ourselves. *Dasein* is a characteristically human form of the *lived future* and inextricably linked to futurity.

Discounting the future

Refers to the economic standpoint of the present and is associated with the assumption that the value of a good diminishes with the passage of time. When applied to environmental problems, for example, it means that temporal distance reduces potential hazards to insignificance, e.g. potential environmental damage costing $1,000,000 in one hundred years time, discounted at a rate of 10% is reduced to a cost of $73 of hazards for the present. See also *Commodification of the future* and *Empty future.*

Disembedding futures

The process whereby specific social contexts and personal biographies are emptied of content and abstracted from their unique historical location. The practice arises with the advent of discourses of progress, economic reasoning and the belief in the technological transformability of the world. See also *Empty future, Commodification of the future* and *Frontier spirit.*

Divination

Practices of telling the future which rely on experts (such as oracles and seers) who read signs within the present to trace there pre-given patterns of the future. The key assumption of such practices is that the future already exists. This assumption is gradually discarded once cultures adopt *knowledge practices* that view the future as subject to human transformation. See also *Open future, Progress* and *Prophecy.*

Empty future

A pervasive *habit of mind* which regards the future as entirely void (apart from the results of our actions), thus essentially ours to fill with content and transform whilst approaching it solely in terms of its relationship to present interests. The idea that unrestricted economic growth is a primary good is perhaps the most influential contemporary example of the dominance of empty futures. See also *Commodification of the future* and *Discounting the future.*

Facta vs futura

A Latin distinction established by de Jouvenel (1967) for the social study of the future. It assumes the future to be *open. Facta* are past events and material things whose influence on the present is the concern of scientists and historians. *Futura* are possibilities that have not yet come about, things and events that will become *facta* only after they have occurred. The distinction implies that facts are material thus real while the future is immaterial, its primary domain being the human mind.

Foresight

A method of future studies designed to establish what possibilities are closer to realisation for a given actor (group, corporation, society) than others. It assumes that the future is more or less open to influence from the present, and that a degree of prediction is possible.

Frontier spirit

Adapted from Jeremy Rifkin's (1994) term 'frontier mentality', it refers to social practices of future making which demarcate, colonise and transform unutilised territories (wilderness) into economically profitable resources. A key characteristic is its strong future orientation and a break with the past that severs established chains of obligation. See also *Disembedding futures.*

Future present

As a standpoint it positions us with reference to deeds and processes already on the way and allows us to accompany actions to their potential impacts on future generations. It enables us to know ourselves as responsible for our *timeprint* and the *time-space distantiated* effects of our actions and inactions. To take that standpoint, however, requires that we first understand invisible process futures in progress as real and *living*. See also *Present futures* and *Present futures vs future presents.*

Futures in the making

Actions that have not yet materialised into symptoms. Although not yet congealed into fixed, empirical facts (thus inaccessible to the usual methods of scientific investigation), they are nonetheless in progress. They are in the process of working themselves out, implying that futurity is ongoing and inescapably incomplete. As processes they are real (effecting reality or *Wirkwelt*) despite not being material in the conventional sense (phenomenal world or *Merkwelt*). See also *Latency, Merkwelt vs Wirkwelt, Natura naturans vs natura naturata.*

Futurology (Futures Studies, Futurism)

The study of futures which is approached from the standpoint of the present and works with de Jouvenel's (1967) distinction between *facta* and *futura*. It assumes the future to be open to human transformation and makes methodological distinctions between the study of probable, possible, and preferable futures.

Habits of mind

Unquestioned, implicit assumptions that form the basis of *knowledge practices* and thus help to construct social environments, making some aspects of the world visible and tangible while neglecting others which are thereby rendered invisible.

Instantaneity

The capacity to compress duration to zero and communicate in *real time*, that is, in the present now. Where previously instantaneity was reserved for face-to-face communication because the movement across space took time, electronic modes of communication and most specifically ICTs have severed the link between time and space and reduced waiting times to close to zero, irrespective of the number of participants involved and distances to be covered. See also *Simultaneity*.

Knowledge practice

Stresses the performative nature of knowledge, the way it constructs its objects and helps to create them. It emphasises the active and constitutive side of knowledge and conveys our belief that transformed understanding and new knowledge change our action potential.

Latency

Refers to futures which are 'on the way' but have not yet materialised, thus cannot be predicted or prophesied with any degree of certainty. Pregnant with the future it directs attention to *future presents* which are real despite not yet having congealed into empirical form. See also *Futures in the making, Merkwelt vs Wirkwelt, Natura naturata vs natura naturans, Phenomenal vs effecting reality*, and *Process vs product.*

Lived future

The way humans and other living entities experience their world as something in the process of being made, anticipate its changing form and participate in its production. See also *conatus*. Organisms adjust and adapt to the potentials present within their environments while humans further involve themselves emotionally, imaginatively and cognitively with the near and distant future, thus extending themselves through *care* from *present futures* into *future presents*. See also *Living future*.

Living future

Refers to halos of evolutionary potential that surround and permeate individuals and make it possible for them to transform others and be transformed by them in surprising and unintended ways. It is neither preformed and pre-determined nor fully indeterminate, empty and open to endless transformation. Rather, it is an embedded future which possesses the still-to-be-determined character of collective *futures in the making*.

Memory of the future

A perspective on the past that views it not as a collection of historical facts but rather as predecessors' images, plans, visions, ambitions and concerns for the future, which either came to fruition or remained unrealised.

Merkwelt vs Wirkwelt

Introduced for (socio-)biological analysis by Jakob von Uexküll and Georg Kriszat (1983/1934), these concepts distinguish between a world that is accessible to our senses (the *phenomenal* world), and a world of processes and forces that is the source of visible outcomes (the '*effecting*' world). See also *Latency* and *Futures in the making*.

Natura naturata vs natura naturans

Two viewpoints on nature, for which Spinoza's *Ethics* (1992/1677) is often cited as a source. *Natura naturata* is the Latin term for the world considered as a set of natural products that have congealed into stable and definite forms, whereas *natura naturans* refers to an ensemble of ongoing processes and forces that produces these products. See also *Futures in the making, Latency*, and *Merkwelt vs Wirkwelt.*

Non-reciprocal responsibility

Responsibility which does not imply reciprocity, such as the duty to care for a dependent. Responsibilities to the future that derive from our ownership of the consequences of our actions are necessarily non-reciprocal, given that those yet to be born are in no position to reciprocate.

Open future

The contemporary future is no longer assumed to be predestined but subject to human shaping and transformation. As such it has to be actively chosen and produced which goes hand-in-hand with an affirmation of collective responsibility for the future. See also *Perfectibility* and *Progress.*

Perfectibility

The doctrine that human beings can transform themselves through the emulation of the divine or knowledge of perfect eternal forms. Introduced in Ancient Greek Orphic religion and, through the cultural influence of neo-Platonism, it became a central assumption of natural philosophy and early social science. It marks a decisive shift away from the belief in pre-given providential futures and towards a future that becomes increasingly open to human influence and effort. See also *Open future.*

Phenomenal reality vs effecting reality

Translations of the German terms *Merkwelt* and *Wirkwelt,* designed to distinguish between reality as a collection of spatio-temporally bounded material facts available to us through our senses, and reality as the ongoing processes which generate these facts and which extend beyond them, in many cases out into the distant future. See also *futures in the making, latency, Merkwelt vs Wirkwelt, natura naturans vs natura naturata,* and *product vs process understanding.*

Politics of posterity

Political decisions that have the capacity to create major consequences that will affect countless future generations who are without voice or vote. As such it opens up questions about what forms of institutions can be legitimate in the light of such extended responsibilities for which there is no political mandate.

Present future

Refers to approaches to the future from the standpoint of the present through which we seek to predict, transform and control the future for the benefit of the present. It projects the future as a terrain that is *empty, open* and subject to colonisation. From this standpoint the factual present is real while *future presents, latent futures* and *futures in the making* lack reality status.

Present future vs future present

Introduced to social science by Niklas Luhmann (1982: 281) who suggested that the present future is rooted in a utopian approach which allows for prediction whilst the future present is technologically constituted and as such enables us to transform future presents into present presents. The distinction plays an important role throughout this book, but is developed here in a different direction. We show that the divergent standpoints involved affect not just our action but also our ethical potential: responsibility for the future requires that we are able to take the standpoint of the future present and have the capacity to move knowledgeably between the two approaches.

Probabilistic prediction

Methods of predicting the future based on statistical evidence which are focused not on individuals and unique events but on aggregates and collectives of these. It forms the basis of early social scientific approaches to the future and has been honed, for example, by economists for the study of long-term trends in the hope of predicting cycles and other patterns of economic change.

Product vs processes understanding

The world we live in and create can be understood in two divergent but ultimately mutually implicated ways: as product and as process. We have access to products of social and natural processes through our senses. They are composed of spatially bounded matter, and can be understood through quantification. The processes that produce them are, by contrast, invisible to us. They encompass the temporal dimension of their products, extending into the past out of which they arose and into their futures. They are intrinsically incomplete and ongoing, and are recognisable only retrospectively. See also: *futures in the making, latency, Merkwelt vs Wirkwelt, natura naturans vs natura naturata*, and *phenomenal reality vs effecting reality.*

Progress

An approach that views the future as arising from actions in the present (rather than being pre-given) and tied to the creation of *present futures*, the aim being the production of happiness through the control of nature in a world of pure potential that is subject to human design and the conscious application of technical knowledge. An unintended but inescapable consequence of the change from *providence* to progress is the contemporary increase in uncertainty and indeterminacy to previously unknown heights.

Promethean power

The capacity, evident in contemporary *knowledge practices* in science, technology, politics and economics, to construct and produce futures without being able to know and take responsibility for the consequences of this future-transforming and traversing power. See also *Structural irresponsibility.*

Prophecy

Telling the future on the basis of divine or other-worldly inspiration where experts act as media or mediators. It assumes supernatural ownership of a future that is pre-given, produced through the workings of supernatural agencies or fate. See also *Divination*.

Providence

The workings of a supernatural agency that is conceived of as owner of the future and believed to exercise overall care for events, secured by a promise or covenant that brings a people (or humanity as such) into the sphere of divine concern. When human beings begin to assume ownership of the future and start to shape it to their design, the belief in providence tends to be deposed from its dominant cultural position and displaced by the pursuit of *progress*.

Real-time action

Social action viewed as taking place in the present, unaffected by the spatial distances involved and made possible by advanced communications technologies. See also *Instantaneity* and *Simultaneity*.

Scenario planning

A plausible description of possible futures, based on coherent and internally consistent sets of assumptions about key relationships and driving forces (e.g. new technological developments, CO_2 emissions, prices). Scenarios are not predictions of what *will* happen; rather they enable the exploration of possible, probable and preferable futures.

Simultaneity

Is action at the same time and refers to the creation of a shared present irrespective of the number of people and the spatial distances involved which is vastly enhanced by the development of information technologies. 9/11 was an event where people with access to a television could share in the global present of this event as it was unfolding in *real time*. See also *Instantaneity*.

Structural irresponsibility

Is constituted at the institutional level of modern industrial societies. In globally networked relations it is produced where the pursuit of *progress* and innovation creates ever greater *timeprints* marked by fundamental uncertainty and indeterminacy while the key institutions (the economy, politics and law) make responsibility dependent on knowledge. In addition, long-term policies routinely pursued by contemporary liberal democracies transgress the temporal boundaries of their political mandates and realms of jurisdiction. In such cases the resulting impacts and costs are in effect externalised to other nations and successor generations, thereby moving the problems outside the institutional sphere of responsibility.

Timeprint

Emphasises the temporal reach of actions without neglecting space and matter. It draws attention to the way that certain *knowledge practices* lead to a consumption of future potential, or appropriation of successors' futures. It alerts us to the problematic relation whereby current future-making extends far beyond any capacity to match our concern and responsibility to the temporal reach of our actions. As such it is the temporal equivalent of the 'ecological footprint', which is based on space and matter, and refers to a measure of demand on nature and compares human consumption of natural resources with the earth's ecological capacity to regenerate them. See also *Structural irresponsibility.*

Time-space distantiation

A term introduced during the early 1980s by the sociologist Anthony Giddens (e.g. Giddens 1984) for the analysis of social modernity which highlights the tendency of actions to have their effects stretched ever further across time and space. It covers both intended and unintended consequences which, due to their global interdependency, can rarely be traced back to their originating actions.

Trading the future/futures trading

Are specific economic practices based on the idea that a promise to buy or sell a commodity at a particular price and specified future date can serve as a means of producing profit and warding off risk. We use the terms in the additional sense that one empty future is assumed to be exchangeable for another, which contrasts fundamentally with embodied, embedded and contextual futures which attach to unique beings and events and are thus unsuitable for abstract exchange. To assume *empty futures* means that any one of a range of possible futures can be realised, depending on what we in the present desire. See also *Abstract future.*

Utopia

An imagined future state of perfection, attained by *progress*, that overcomes current limitations on human potential. Images of utopian societies became a major theme of European literature from the 1500s onwards, being initially located in geographically distant places, but later transferred to the temporal realm of near or far futures. Alongside *perfectibility* and *progress*, it marked another point of departure from the belief in *providence* and predestination. See also *Present future vs future present.*

Valorisation of speed

Speed provides evolutionary, economic and military advantages. Economically, the speed of achieving a given financial return is a variable in determining efficiency and profitability. Environmentally it is a variable in energy consumption and pollution. In contemporary contexts where the *timeprint* of social practices steadily increases, the valorisation of speed narrows the temporal focus to the present, thereby decreasing our capacity to take a long-term perspective and accompany actions to their potential *time-space distantiated* effects.

BIBLIOGRAPHY

Adam, B. 1990. *Time and Social Theory.* Cambridge: Polity Press.

——. 1998. *Timescapes of Modernity: The Environment and Invisible Hazards.* London: Routledge.

——. 2000a. 'Mediated Risk: BSE in the Broadsheets', in Allan, S., Adam, B. and Carter, C., eds., *Environmental Risks and the Media,* London and New York: Routledge, pp. 117–129.

——. 2000b. 'The Temporal Gaze: Challenge for Social Theory in the Context of GM Food', Millennium Issue *British Journal of Sociology,* 51/1:125–142.

——. 2001. 'The Value of Time in Transport' in Giorgi, L. and Pohoryles, R. eds. *Transport Policy and Research: What Future?* Aldershot: Ashgate, 130–143.

——. 2004. *Time.* Cambridge and Malden, MA: Polity.

——. 2005. 'Futurity from a Complexity Perspective'. http://www.cardiff.ac.uk/socsi/futures/wp_ba_complexity230905.pdf.

Alexander, J. 1992. *The Astrological Manager. A New Approach to Business, Success and Destiny.* New York: Annes.

Anderson, Walter Truett. 1987. *To Govern Evolution.* Boston: Harcourt.

Arendt, H. 1998/1958. *The Human Condition.* Chicago: Chicago University Press.

Assmann, J. 2000. *Der Tod als Thema der Kulturtheorie.* Frankfurt a.M.: Editions Suhrkamp.

——. 2001/1984. *The Search for God in Ancient Egypt.* Trans. David Lorton. Ithaca & London: Cornell University Press.

Aveni, A.F. ed. 1975. *Archaeoastronomy in Pre-Columbian America.* Austin: University of Texas Press.

Bauman, Z. 2000. *Liquid Modernity.* Cambridge and Malden, MA: Polity Press.

Beck, U. 1992/1986. Risk Society: Towards a New Modernity. London: Sage.

——. 1999. *World Risk Society.* Cambridge: Polity.

Becker, E. 1973. *The Denial of Death.* New York: Macmillan.

Beinhocker, E.D. 2006. *The Origin of Wealth: Evolution, Complexity, and the Radical Remaking of Economics.* Cambridge, MA: Harvard Business School Press.

Bell, W. 2003/1997. *Foundations of Futures Studies. History, Purposes, and Knowledge.* 2 vols. New Brunswick: Transaction Publishers.

Bell, W. and Mau, J. eds., 1971. *The Sociology of the Future: Theory, Cases, and Annotated Bibliography,* New York: Russell Sage Foundation.

Benjamin, J. 1988. *The Bonds of Love: Psychoanalysis, Feminism and the Problem of Domination.* New York: Pantheon Books.

Berman, M. 1983. *All that is Solid Melts into Air: The Experience of Modernity.* London: Verso.

Bertman, S. 1998. *Hyperculture. The Human Cost of Speed.* Westport, CT: Praeger.

Biesecker, A. 1998. 'Economic Rationales and a Wealth of Time,' *Time & Society* 71: 75–91.

Birnbacher, D. 2001. 'Philosophical foundations of responsibility', in Auhagen, A.E. and Bierhoff, H.-W. Eds., *Responsibility: The many faces of a social phenomenon.* ed., London: Routledge, 9–22.

Blumenberg, H. 1986. *Lebenszeit und Weltzeit.* Frankfurt a.M.: Surhkamp.

Boden, D. 2000. 'Worlds in Action: Information, Instantaneity and Global Futures Trading', in Adam, B., Beck, U., v. Loon, J. eds., *The Risk Society and Beyond.* London: Sage, 183–197.

Bohm, D. 1983. *Wholeness and the Implicate Order.* London: ARK.

Bourke, V.J. (ed.) 1983. *The Essential Augustine.* Indianapolis: Hackett Publishing Company.
Briggs, J. and Peat, D. 1989. *Turbulent Mirror.* New York: Harper and Row.
Brög, W. 1996. *Presentation to Car Free City Conference, May 1996.* Copenhagen: Commission of the European Communities.
Brown, N., Rappert, B. and Webster, A. 2000. 'Introducing Contested Futures. From "Looking into" the Future to "Looking at" the Future', in Brown, N., Rappert, B. and Webster, A. eds. *Contested Futures. A Sociology of Prospective Techno-science.* Aldershot: Ashgate, 3–20.
Brumbaugh, R.S. 1966. Applied Metaphysics: truth and passing time. *Review of Metaphysics* 19: 647–666.
Burtt, E.A. 1959. *The Metaphysical Foundations of Modern Physical Science.* London: Routledge and Kegan Paul.
Bury, J.B. 1955/1932. *The Idea of Progress. An Inquiry into its Growth and Origin.* New York: Dover Publications Inc.
Butterfield, H. 1965. *The Origins of Modern Science 1300–1800.* London: G. Bell & Sons.
Byrne, D. 1998. *Complexity Theory and the Social Sciences.* New York/London: Routledge.
Capra, F. 1996. *The Web of Life.* New York: Anchor/Doubleday.
——. 2003. *The Hidden Connections.* London: Flamingo.
Chisholm, R. 1967. He Could Have Done Otherwise. *The Journal of Philosophy,* 64 (13), 409–417.
Cohen, J. 1964. *Behaviour in Uncertainty.* London: Allen & Unwin.
Colborn, T., Meyers, J.P. and Dumanoski, D. 1996. *Our Stolen Future. How Manmade Chemicals are Threatening our Fertility, Intelligence and Survival.* Boston: Little, Brown & Company.
De Jouvenel, B. 1967. *The Art of Conjecture.* Trans. N. Lary, London: Weidenfeld and Nicolson.
Deleuze, G. & Guattari, F. 1988/1981. *A Thousand Plateaus.* Trans. Brian Massumi. London: Athlone.
Deleuze, G. 1994/1968. *Difference and Repetition.* Trans. Paul Patton. London: Athlone.
Drury, N. 2000. *Shamanism. An Introductory Guide to Living in Harmony with Nature.* Boston: Elements.
Eagle Alliance, http://www.eaglealliance.org/, http://www.vanderbilt.edu/radsafe/9611/msg00220.html.
Ecologist. 'Nuclear Power Dossier—Building a Nuclear Power Station'. June 2006. 41–57.
Ekelund, R.B. and Hébert, R.F. 1997. *A History of Economic Theory and Method.* New York: McGraw-Hill.
Eliade, M. 1989/1959. *Cosmos and History: The Myth of the Eternal Return.* London: Arcana.
Eliot, T.S. 1963. *Collected Poems 1909–1962.* London: Faber and Faber.
Evans, R.J. 1997. Soothsaying or Science: Falsification, Uncertainty and Social Change in Macro-econometric Modelling, *Social Studies of Sciences,* 273: 395–438.
Ferguson, D. 2000. *The History of Myths Retold.* London: Chancellor Press.
Flannery, T.F. 1994. *The Future Eaters. An Ecological History of the Australian Lands and People.* Sidney: Reed Books.
Frankfurt, H. 1982. The Importance of What We Care About. *Synthèse* 53(2), pp. 257–272.
Friedman, M. 1993. *What Are Friends For?: Feminist Perspectives on Personal Relationships and Moral Theory.* Ithaca, NY: Cornell University Press.
Gadamer, H.-G. 1994/1960. *Truth and Method.* Trans. Joel Weisheimer. London: Sheed & Ward.

Geddes & Grosset 1997. *Ancient Egypt. Myth and History.* New Lanark, Scotland: Gresham Publishing Company.
George, S. 1989. *A Fate Worse than Debt.* Harmondsworth: Penguin.
——. 1992. *The Debt Boomerang.* London: Pluto Press.
Giddens, A. 1984. *The Constitution of Society.* London: Polity Press.
Gilligan, C. 1982. *In a Different Voice.* Cambridge, Mass.: Harvard University Press.
Goodwin, B., and Keith Taylor. 1982. *The Politics of Utopia: A Study in Theory and Practice.* London: Hutchinson.
Gribbin, J. 2005. *Deep Simplicity: Chaos Complexity and the Emergence of Life.* Harmondsworth: Penguin Press Science.
Griffiths, J. 1999. *Pip Pip: A Sideways Look at Time.* London: Harper Collins.
Haack, S. 2005. Trial and Error: The Supreme Court's Philosophy of Science. *American Journal Of Public Health* 95 (Supplement 1), pp. 66–73.
Hassan, R. 2003. Real-time and the New Knowledge Epoch. *Time & Society*, 122/3: 225–41.
Hayles, K.N. 1990. *Chaos Bound. Orderly Disorder in Contemporary Literature and Science.* Ithaca: Cornell UP.
Hayward, R. 1997. "From the Millennial Future to the Unconscious Past: The Transformation of Prophecy in Early Twentieth-Century Britain", in Thaite, B. and Thornton, T. eds., *Prophecy*, Stroud: Sutton Publ. Ltd. pp. 161–180.
Heggie, D. ed. 1982. *Archaeoastronomy in the Old World.* Cambridge: Cambridge University Press.
Heidegger, M. 1988/1927. *Being and Time.* Oxford: Blackwell.
Henderson, Hazel. 1988. *The Politics of the Solar Age: Alternatives to Economics.* Knowledge Systems Inc.
Hillman, M. & Plowden, S. 1996. *Speed Control and Transport Policy.* London: Policy Studies Institute.
Hoekema, A.A. 1994/1979. *The Bible and the Future.* Grand Rapids, Michigan: William B Eerdmans Publ. Co.
Hornung, E. 1999/1997. *The Ancient Egyptian Books of the Afterlife.* Trans. David Lorton. Ithaca & London: Cornell University Press.
Hughes, J. 2006. "Nuclear Power Dossier—Building a Nuclear Power Station", *The Ecologist*, June 2006.
Inayatullah, S. 2005. *Questioning the Future.* Tamsui and Taipei, Tamkang University.
Ingarden, R. 1970. *Über die Verantwortung.* Stuttgart: Reclam.
Jacobs, M. 1991. *The Green Economy. Environment, Sustainable Development and the Politics of the Future.* London: Pluto Press.
Johnson, W.E. 1921. *Logic.* Cambridge: University Press.
Jonas, H. 1982/1968. *The Phenomenon of Life: Towards a Philosophical Biology.* Chicago; London: University of Chicago Press.
——. 1984/1976. *The Imperative of Responsibility.* Chicago; London: University of Chicago Press.
Kant, I. 1963/1784. 'Idea for a Universal History from a Cosmopolitan Point of View', in Lewis White Beck, ed. *On History.* Indianapolis: Bobbs-Merrill.
——. 1993/1785. *Groundwork of the Metaphysics of Morals*, trans. J.W. Ellington. Indianapolis: Hackett.
Kearnes, M., McNaghten, P. and Wilson, J. 2006. *Governing at the Nano Scale: People, Policies and Emerging Technologies.* London: Demos.
Kern, S. 1983. *The Culture of Time and Space 1880–1918.* London: Weidenfeld and Nicolson.
King, B. 2000. *Runes. An Introductory Guide to Interpreting the Ancient Wisdom of the Runes.* Boston: Element.
Kohlberg, L. 1981. *Essays on Moral Development.* London: Harper and Row.

Koppe, J.G., and Jane Keys. 2001. PCBs and the Precautionary Principle. In: Harremöes, P. et al. eds. *Late Lessons from Early Warnings: The Precautionary Principle 1896–2000.* Vol. 22. Copenhagen: European Environment Agency, pp. 64–75.

Korsgaard, C. 1996. *Creating the Kingdom of Ends.* Cambridge & New York: Cambridge University Press, 1996.

Lakoff, G. and Johnson, M. 2003/1980. *Metaphors We Live By.* Chicago: University of Chicago Press.

Latour, B. 1993. *We Have Never Been Modern.* Trans. Catherine Porter. New York: Harvester Wheatsheaf.

——. 2004. *The Politics of Nature.* Trans. Catherine Porter. Harvard: Harvard UP.

Law, J. and Mol, A. 2002. *Complexities: Social Studies of Knowledge Practices.* Durham, NC: Duke University Press.

Le Goff, J. 1980. *Time, Work and Culture in the Middle Ages.* Chicago: University of Chicago Press.

Lewinsohn, R. 1961. *Science, Prophecy and Prediction.* Trans. Arnold J. Pomerans, New York: Bell Publishing Company, Inc.

Lidsky, L.E. 1983. "The Trouble with Fusion". *Technology Review* 86: 32–44.

Ling, T. 2000. 'Contested Health Futures', in Brown, N., Rappert, B. and Webster, A. eds. *Contested Futures. A Sociology of Prospective Techno-science.* Aldershot: Ashgate, pp. 251–70.

Lippincott, K., Eco, U., Gombrich, E.H., et al. 1999. *The Story of Time.* London: Merrell Holberton.

Littleton, S.C., ed. 2002. *Mythology. The Illustrated Anthology of World Myth and Storytelling.* London: Duncan Baird.

Locke, J. 1988/1689. *Two Treatises of Government.* Cambridge: Cambridge University Press.

Long Now Foundation, http://www.longnow.org.

Luhmann, N. 1982. *The Differentiation of Society.* New York: Columbia University.

MacLean, D. 1983. "A Moral Requirement for Energy Policies" in MacLean, D. and Brown, P.G. eds. *Energy and the Future.* Totowa, NJ: Rowman and Littlefield, pp. 180–196.

Malinowski, B. 1920. Kula; the Circulating Exchange of Valuables in the Archipelagoes of Eastern New Guinea. *Man* 51: 97–105.

——. 1922 *Argonauts of the Western Pacific.* London: Routledge & Kegan Paul.

Mannheim, K. 1936. *Ideology and Utopia.* London: Routledge and Kegan Paul.

Marx, K. 1977/1856. "Speech at the Anniversary of the People's Paper", in McLellan, D., *Karl Marx. Selected Writings.* Oxford: Oxford University Press, 338–339.

Marx, K., and Engels, F. 1967/1848. *The Communist Manifesto,* Harmondsworth. Penguin.

Max-Neef, M. 1992. "Development and Human Needs", in Ekins, P. and Max-Neef, M. eds. *Real Life Economics.* London: Routledge, pp. 197–214.

May, Graham H. 1996. *The Future is Ours: Foreseeing, Managing and Creating the Future.* Westport, CT: Praeger Studies on the 21st Century.

Mbiti, J.S. 1985/1969. *African Religions and Philosophy.* London: Heinemann.

McLuhan, M. 1973/1964. *Understanding Media.* London: Routledge and Kegan Paul.

Moltmann, J. 1967. *Theology of Hope.* Trans. J. W Leitsch. New York: Harper and Row.

Morgan, O. 2006. Nuclear Costs to Hit Historical 90 Billion Warns Brown. *Observer* 4 June 2006.

Montague, P. 2005. "Fusion Illusion", *New Internationalist,* 382: 14–15.

More, Thomas. 2003/1516. *Utopia.* Harmondsworth: Penguin.

Nelis, A. 2000. "Genetics and Uncertainty", in Brown, N., Rappert, B. and Webster, A. eds. *Contested Futures. A Sociology of Prospective Techno-science.* Aldershot: Ashgate, pp. 209–28.

Nietzsche, F.W. 1966/1882. *Beyond Good and Evil.* Trans. W. Kaufmann. New York: Vintage.

Nowotny, H. 1994/1989. *Time: The Modern and Postmodern Experience.* Trans. N. Plaice. Cambridge and Cambridge, MA: Polity.
——. 2001. *Coping with Complexity. On Emergent Interfaces between the Natural Sciences, Humanities and Social Sciences.* Manuscript.
Nussbaum, M.C. 1990. *Love's Knowledge: Essays on Philosophy and Literature.* New York; Oxford: Oxford University Press.
O'Neill, J. 1993. *Ecology, Policy and Politics.* London; New York: Routledge.
Ovitt Jr., G. 1987. *The Restoration of Perfection: Labor and Technology in Medieval Culture.* New Brunswick, London: Rutgers University Press.
Passmore, J. 2000. *The Perfectibility of Man.* Indianapolis: Liberty Fund.
Peccei, A. 1982. *One Hundred Pages for the Future: Reflections of the President of the Club of Rome.* London: Futura.
Pellizzoni, L. 2004. Responsibility and Environmental Governance. *Environmental Politics* 13(3), pp. 541–565.
Pimentel, D. 1993. *World Soil Erosion and Conservation.* Cambridge.
Pimentel, D. et al. 1995. Environmental and Economic Costs of Soil Erosion and Conservation Benefits. *Science* 267: 1117–1123.
Plotinus 1966/253–270. *Enneads.* London: Heinemann.
Polanyi, K. 1971. *Primitive, Archaic and Modern Economies.* Boston: Beacon Press.
Prigogine, I. and Stengers, I. 1984. *Order out of Chaos: Man's New Dialogue with Nature.* London: Heinemann.
Protevi, J. 2001. *Political Physics.* London: Athlone Press.
Reanney, D. 1995. *The Death of Forever: A New Future for Human Consciousness.* London: Souvenir Press.
Rifkin, J. 1987. *Time Wars.* New York: Henry Holt.
——. 1991. *Biosphere Politics.* San Francisco: Harper Collins.
——. 1994. *Beyond Beef: The Rise and Fall of the Cattle Culture,* London: Thorsons/ Harper Collins.
Rosetta Project, http://www.rosettaproject.org.
Routh, G. 1975. *The Origin of Economic Ideas.* London: Macmillan Press.
Ruggles, C.L.N. 1994. "British Archaeoastronomy", in S.L. Macey ed., *Encyclopedia of Time.* New York: Garland Publishing, 68–69.
Sabelis, I.H.J. 2001. *Managers' Times. A Study of Times in the Work and Life of Top Managers.* Amsterdam: Vrije Universiteit.
Samdup, Lama Kazi Dawa. 1957. *The Tibetan Book of the Dead.* Oxford: Oxford UP.
Sartre, J.-P. 2003/1943. *Being and Nothingness.* London/New York: Routledge Classics.
Serres, M. 1999/1982. *Genesis.* Michigan: University of Michigan Press.
Shaw, E. 1997. *The Wordsworth Book of Divining the Future.* Ware: Wordsworth Reference.
Shrader-Frechette, K. 1993. *Burying Uncertainty: Risk and the Case Against Geological Disposal of Nuclear Waste.* Berkeley: University of California Press.
Simmel, G. 1982/1900. *The Philosophy of Money.* Trans. Tom Bottomore and David Frisby. London: Routledge and Kegan Paul.
Slaughter, Richard A. 1995. *The Foresight Principle.* Greenwood Press.
Smith, A. 1975/1776. *The Wealth of Nations.* London: J.M. Dent & Sons.
Spinoza, B. 1992/1677. *Ethics.* Trans. Samuel Shirley. Indianapolis: Hackett.
Steiner, H. 1983. "The Rights of Future Generations", in MacLean, D. and Brown, P.G. eds. *Energy and the Future.* Totowa, NJ: Rowman and Littlefield, pp. 151–165.
Stern, D. 1985. *The Interpersonal World of the Infant.* New York: Basic Books.
Stone, Christopher D. 1996. *Should Trees Have Standing?* Oxford: Oxford University Press.
Storm van Leeuwen, J.W. 2005. 'A Nuclear Power Primer'. http://www.opendemocracy.net/globalization-climate_change_debate/2587.jsp.
Sullivan, K. 1998. *This New Promethean Fire: Radioactive Monsters and Sustainable Nuclear Futures.* PhD Thesis, University of Lancaster.

Thaite, B. and Thornton, T. 1997. "Then Language of History: Past and Future in Prophecy", in Thaite, B. and Thornton, T. eds., *Prophecy.* Stroud: Sutton Publ. Ltd. pp. 1–16.
Thrift, N. 1999. The Place of Complexity. *Theory, Culture & Society*, 16/3: 31–70.
Todorov, T. 1996. *Facing the Extreme: Moral Life in the Concentration Camps.* London: Weidenfeld & Nicholson.
Uexküll, J.V. and Kriszat, G. 1983/1934. *Streifzüge durch die Umwelten von Tieren und Menschen.* Frankfurt, a.M.: Fischer.
UK Government Interdepartmental Liaison Group on Risk Assessment (ILGRA) 2002. *The Precautionary Principle: Policy and Application.* Available at: http://www.hse.gov.uk/aboutus/meetings/ilgra/pppa.pdf [Accessed: 13/12/06].
Urry, J. 2003. *Global Complexity.* Cambridge: Polity.
Virilio, P. 1991. *La Vitesse*, Paris: Éditions Flammarion.
——. 1995/1993. *The Art of the Motor.* Trans. J. Rose, Minneapolis: University of Minnesota Press.
——. 1997. *Open Sky.* Trans. Julie Rose. London: Verso.
——. 2000/1999. *The Information Bomb.* Trans. Chris Turner. London: Verso.
Waskow, A.I. 1969. 'Looking Forward: 1999', in Jungk, R. and Galtung, J. eds., *Mankind 2000.* Oslo: Universitets-forlaget, pp. 78–98.
Weber, M. 1989/1904–5. *The Protestant Ethic and the Spirit of Capitalism.* Trans. T. Parsons. London: Unwin Hyman.
——. 1969/1919. 'Science as a Vocation', in Gerth, H.H. and Mills, C. Wright eds., *From Max Weber: Essays in Sociology.* London: Routledge and Kegan Paul, pp. 129–58.
Weick, K.E. and Sutcliffe, K.M. 2001. *Managing the Unexpected: Assuring High Performance in an Age of Complexity.* Hoboken, NJ: Jossey Bass Wiley.
Wendorff, R. 1991. *Die Zeit mit der wir leben.* Herne: Heitkamp.
Whitehead, A.N. 1929. *Process and Reality.* New York: Harper and Row.
Whitelegg, J. 1993. *Transport for a Sustainable Future. The Case for Europe.* London: Belhaven Press.
——. 1997. *Critical Mass. Transport, Environment and Society in the Twenty-first Century.* London: Pluto Press.
Wolin, R. 2001. *Heidegger's Children: Hannah Arendt, Karl Löwith, Hans Jonas, and Herbert Marcuse.* Princeton, NJ: Princeton University Press.
Wood, J. 2000. *The Celtic Book of Living and Dying.* London: Duncan Baird Publ.
Wynne, B. 2005. Reflexing Complexity: Post-Genomic Knowledge and Reductionist Returns in Public Science. *Theory, Culture and Society* 22: 67–94.
Young, I.M. 2006. Responsibility and Global Justice: A Social Connection Model. *Social Philosophy and Policy* 23(1), pp. 102–130.
Zilsel, E. 1941–42. The Sociological Roots of Science. *American Journal of Sociology* 47, pp. 544–562.
Zohar, D. 1983. *Through the Time Barrier*, London: Paladin.

NAME INDEX

SUBJECT INDEX